Sloth

Alan Rauch

REAKTION BOOKS

In memory of Riva Rauch, a source of love and energy

Published by
REAKTION BOOKS LTD
Unit 32, Waterside
44–48 Wharf Road
London N1 7UX, UK
www.reaktionbooks.co.uk

First published 2023
Copyright © Alan Rauch 2023

All rights reserved

No part of this publication may be reproduced, stored in a retrieval
system or transmitted, in any form or by any means, electronic,
mechanical, photocopying, recording or otherwise, without the prior
permission of the publishers

Printed and bound in India by Replika Press Pvt. Ltd

A catalogue record for this book is available from the British Library

ISBN 978 1 78914 799 5

Contents

Preface

How sweet to be a sloth.
Michael Flanders and Donald Swann, 'The Sloth' (1962)

It is tempting to say at the very outset that sloths need no intro-
duction. But of course, not only would that preclude the very
purpose of this book but, more importantly, it would simply be
inaccurate. Our familiarity with sloths in contemporary Western
culture is understandably limited to a set of familiar facts: first,
that sloths are slow; second, that they hang upside-down; and
third, that they are cute. All of these commonplaces may be true,
but they do little justice to the sloths in the modern era, much less
to their massive ancestors, the giant ground sloths, many of
which roamed through the Americas – North, Central *and* South
– as recently as 6,000 years ago.

Modern-day tree sloths and ancient ground sloths are equally
fascinating, but they are so different in physical appearance and
so separated by time that it is difficult to consider both in a uni-
fied manner. Tree sloths are commonplace to the Indigenous
peoples of Central and South America, but their form and behaviour struck European naturalists as bizarre, and scientists have
been puzzled by their structure and function ever since they were
first identified in the sixteenth century. To this day they remain
a puzzle, and as tempting as it is to lump together contemporary
sloths – two-toed and three-toed – they are very different and less
closely related to each other than they are to their prehistoric
ancestors. Even the apparently simple fact, as is widely known,

Sloth relaxing.

7

that sloths have a greenish hue because their slow pace allows algae to grow on them is deceptive. Yes, both genera of sloth are green (advantageous, perhaps, as camouflage), but the very hairs of the two groups that allow algae to grow are (as we shall see) radically different.[1]

The discovery of the equally bizarre giant ground sloths two centuries later challenged zoologists once again, not merely because these sloths were also odd in both form and function, but because the link to modern sloths was, at best, a puzzle in its own right. The giant sloths are, as we shall see, strange indeed, and they also had the distinction of being among the first fossils to be assembled and thus to serve as evidence for a world inhabited by massive, yet now extinct, creatures. Well before dinosaurs dominated our exhibition halls, giant sloths found their way into museums: skeletons of these improbable ancient sloths, with names like *Megatherium*, *Megalonyx* and *Mylodon*, captured the imagination of nineteenth-century scientists, citizens and children. How was it, people wondered, that such large and formidable creatures could exist, and how was it that they could eventually become extinct? But no less pressing a question was how it is that the only living relatives (not to say, in these pre-Darwinian times, 'the descendants') of these massive creatures are the pitiable and seemingly insignificant sloths of South America.

These were not questions to be ignored, and the fact remains that we know much less about sloths, prehistoric and contemporary, than we care to admit. The very origin of modern sloths, rooted deeply in the history of the South American continent as it emerged as a separate land mass, is unclear. No less perplexing is the evolutionary transition of the great sloths of the Palaeocene to the familiar sloths of today. Fossil remains, in both cases, are scarce and can only begin to suggest the physical and behavioural adaptations that developed into these remarkable and highly specialized forms. Still, these animals *were* and *are* remarkable, and we are making great advances, particularly in the last few decades, in understanding them and, no less important, in taking them very seriously.

That said, it is probably best to recognize at the outset of this book that there are three kinds of sloth that will consume our attention: sloths of the past, sloths of the present and sloths of social and cultural imagination. That may seem intuitive, but my point is that sloth narratives are simply more restricted than those of domesticated animals, such as dogs and horses, or animals that have a well-developed ecological and evolutionary history, such as the primates. One might say that part of the charm of sloths is that they are thoroughly engaging and yet also irresistibly enigmatic.

I came to sloths very young and very confused by their puzzling appearance. As a frequent visitor, as both a child and an adult, to McGill University's Redpath Museum in the heart of Montreal, I regularly faced a huge fossil mounted upright against a tree. At first, I assumed it was a dinosaur, but as I took more time (with some parental guidance) to read the signage, I learned that it was a *Megatherium*, that is to say a *giant* sloth. How, I wondered, was I to reconcile the remains of this great beast with the little I knew about the modestly sized sloths of the South American rainforests? As I have already suggested, I am not alone in still not knowing the answer, but this is a question that has sustained me for a lifetime and that has intrigued many accomplished scholars for at least two centuries. And no doubt this was the question that dogged Sir John William Dawson (1820–1899), founder of the Redpath Museum, who was simultaneously Canada's most renowned geologist and staunchest Creationist. But Dawson, who was also principal of what was then McGill College, found comfort, if not answers, in faith. Thus, it is not surprising that visitors to the new Redpath Museum were greeted with a plaque that read, 'O, Lord, how manifold are Thy works! All of them in wisdom Thou hast made.'[2]

I have written elsewhere that part of the purpose of mounting giant sloths as exhibits in the nineteenth century was to prompt questions of faith among museumgoers. Here were these amazing animals, massive and mighty in their time, now reduced to creatures that were more likely to be pitied or ridiculed. The contrast between ancestor and descendent is so extreme that it was impossible not to wonder about it. Robert Chambers (1802–1871), the publisher and amateur geologist, articulated this very perspective in his ambitious (though anonymous) work on transmutation, *Vestiges of the Natural History of Creation* (1844). 'We are . . . surprised', Chambers wrote inquisitively about *Megatheria*, 'by finding such gigantic proportions in an animal called the megatherium, which ranks in an order now assuming much humbler forms – the edentata – to which the sloth, ant-eater, and armadillo belong.'[3] The 'surprise' was puzzling for many natural historians, whether professional or amateur. This was particularly true because the dramatic reduction in size violated what was known as 'Cope's Rule', which asserted that mammals tended to increase in size over geological time. The classic model for this claim, which is attributed to the great American palaeontologist Edward Drinker Cope (1840–1897), is the horse, which evolved in

Cope's Rule.

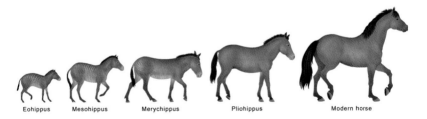

Eohippus   Mesohippus   Merychippus   Pliohippus   Modern horse

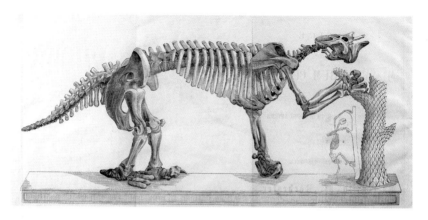

form from the dog-sized *Eohippus* to the modern *Equus*. The conundrum of the giant ground sloth is that it seemed to have endured a great fall – as it were – from its majestic, if not imperious, bulk down to an inconsequential tree-dweller. For biblically literate observers, perhaps thinking of 2 Samuel 1:25–7, the radical transition of form was a simple reminder of how easily the mighty are fallen. And for those such as William Dawson, who were devoted to reconciling faith with science, it was an instructive instance that God's creations did not have to conform to any 'rules' but rather to a divine plan. The inversion of one such rule, by the widely admired Cope, is immediately apparent in the illustration from Henry Augustus Ward's (1834–1906) *Catalogue of Casts of Fossils, from the Principal Museums of Europe and America, with Short Descriptions and Illustrations* (1866), which visually contrasts the tiny ground sloth against its colossal ancestor.[4] No one who contemplated this illustration (and there were many others like it) could resist agreeing with Dawson that the descendants of the great sloths were simply less magnificent than their ancestors. That is not to say, particularly from our present-day perspective, that one ought to reject or disdain the 'diminished' modern sloth.

Illustration showing the contrast between the enormous giant sloth and its diminutive contemporary descendant. From Henry A. Ward, *Catalogue of Casts of Fossils . . .* (1866).

Even Dawson tried to provide a context for understanding unusual, or what might be considered abnormal, animals, whether living or fossilized; we must regard and welcome the Post-Pliocene, Dawson wrote, as a 'garden of the Lord, fitted for the reception of His image'.[5] For many students of animal life, there was no escaping Dawson's theological perspective of life on earth, particularly because he was a powerful enough rhetorician to boldly, if unsuccessfully, assert the presence of the divine in *every* facet of nature.

Needless to say, for Dawson, a *Megatherium* was an absolute must for his new museum, and so he turned to Henry Augustus Ward for a cast of the gigantic sloth. Dawson knew that a *Megatherium* cast had become de rigueur for the very best natural history museums, primarily for its immense size, but equally significant was that it underscored his Creationist views. The opening of the museum – depicted in the *Canadian Illustrated News* – was a soirée to be reckoned with and looming over all of the elegant guests was Ward's *Megatherium*.[6]

Perhaps this is a little too much nostalgia, but I hope it helps to suggest the complexity of tracing the history and biology of the sloth (giant or arboreal), and conceivably of any animal that we care to examine closely. In addition to its biological and ecological status, every animal has meaning as a cultural icon, and that meaning, by necessity, will vary over time as the perception of the creature resonates with new ideas and new concerns. The relaxed and carefree attitude that still informs our perception of modern sloths, for example, is now balanced (or at least *should* be balanced) by our concern for the destruction of their habitat in South America. No less important is the awareness that animals no longer 'belong' to us (if they ever did) and so ought not to be available for commercial petting or social experiences. Whatever

Megatherium skeleton in the Gallery of Palaeontology and Comparative Anatomy, Paris.

pleasure individuals derive from meeting a sloth is necessarily counterbalanced by the intense level of stress that sloths must feel away from their natural environment, not to mention being handled by a parade of strangers.

And I count myself among those strangers, although I should note that throughout the period that I worked on a rescue ranch with sloths, I never once held any of the animals in my arms. I take that as a point of pride, but credit is due to the staff at the facility who enforced a serious hands-off policy. My experience, which included gathering food for the sloths, building an enclosure to aid with transition back to the wild and simply observing the resident animals (some that were housed permanently because of injuries and some that were rehabilitating for eventual transition to the wild), was not diminished by the lack of contact. Their world and the way they experience their environment is so remote from any one of us that it can't help but seem presumptuous to intrude on *their* space.

A pair of juvenile Hoffman's two-toed sloths in a rescue centre.

Reception for the American Association for the Advancement of Science in McGill University's new Redpath Museum, 1882. The *Megatherium* skeleton is in the background.

Megatherium centrally positioned in the Redpath Museum, c. 1910.

Juvenile
three-toed sloth,
Costa Rica.

That doesn't mean that we can't have affection for sloths, ancient or modern, or be fascinated by their unique lives. Nor does it mean that we should not be proactive about their welfare and concerned about the degradation of their habitat. These are all factors that motivated me to write this book and, in doing so, to make sloths meaningful and perhaps even magical in the lives of current readers. We all need to encourage what might be called 'slothful' enthusiasm to support and sustain these wonderful animals in the wild, and to protect the irreplaceable rainforests that support them and countless other species.

Sloths of the Rainforest

PALE-THROATED SLOTH
Scientific Name: Bradypus tridactylus
Length: 20–30" (50–76 cm) (F) / 18–22" (45–55 cm) (M)
Conservation Status: Least Concern

LINNAEUS'S TWO-TOED SLOTH
Scientific Name: Choloepus didactylus
Length: 18–34" (46–86 cm)
Conservation Status: Least Concern

BROWN-THROATED SLOTH
Scientific Name: Bradypus tridactylus
Length: 17–31" (42–80 cm)
Conservation Status: Least Concern

PYGMY THREE-TOED SLOTH
Scientific Name: Bradypus pygmaeus
Length: 19–21" (58–72 cm)
Conservation Status: Critically Endangered

MANED SLOTH
Scientific Name: Bradypus torquatus
Length: 18–20" (45–50 cm)
Conservation Status: Vulnerable

HOFFMANN'S TWO-TOED SLOTH
Scientific Name: Choloepus hoffmanni
Length: 21–28" (54–72 cm)
Conservation Status: Least Concern

© BLACK & SKULL • DIVI-ART INKU

1 The Modern Sloth

As more and more people care, so there is greater hope
that the sloths, along with their forest world and the other
wondrous creatures that live there, will survive.
Jane Goodall[1]

Contemporary sloths are generally referred to as tree sloths, not
only to distinguish them from their massively larger ancestors,
the ground sloths, but because they are predominantly arboreal.
The question of how the huge ancestral ground sloths found their
evolutionary way to becoming smallish animals inhabiting tree
canopies is, as we have noted earlier, both complicated and poorly
understood. But it nevertheless involves fascinating turns of
events – or should I say, 'turns of survival and adaptation' – that
render the history of sloths thoroughly interesting. So, notwithstanding the title of this chapter, it is almost impossible to talk
about modern living sloths without invoking some palaeontology.

The taxonomy of sloths has been a matter of controversy for
a long time, and many questions about 'true' lineages persist.
Currently, sloths are included in the superorder called Xenarthra,
which includes, in addition to tree sloths and now-extinct ground
sloths, both armadillos and anteaters. Two extinct families –
the heavily armoured Pampatheres and the carapace-carrying
Glyptodonts, both ancestrally related to armadillos – were also
included in this assorted group. The Xenarthra in general can be
characterized by the unusual vertebral joints (*xénos* and *árthron*
in Greek) that are not found in other mammals. The joints in question are lumbar vertebrae in the lower back, which connect with
each other in a way that helps support the large pelvis and the tail

Compilation
of all living sloths
in Central and
South America,
by Roger Hall.

structure. This formation was helpful when the sloths raised themselves up to forage from trees or to brace themselves when digging.

Many features of the Xenarthrans are considered primitive, and they may well be among the most ancient mammals. The teeth have none of the complexity of other mammals, and the genitalia of the Xenarthrans, both male and female, are, to say the least, unusual. Males have internal testicles and females have a single genital structure, considered relatively primitive in its formation, that functions as uterus and vagina.

The Xenarthrans, all predominantly herbivores and insectivores, originated in South America and never migrated beyond the Americas. What's more, all of the larger forms (the megafaunal species) became extinct towards the end of the Pleistocene about 11,000 years ago – at least this seems true for North America – while giant sloths may have lasted for another millennium or so in South America, and some island-dwelling species (in the West Indies) may have lived until as recently as 4,500 years ago. In North America, there are some signs that the advent of humans may have contributed to the extinction of the giant sloths, but much of the evidence is circumstantial. Palaeontological work in White Sands, New Mexico, confirms that humans walked in the very footsteps of giant sloths, possibly with predatory intentions. Recently a 'kill site' has been discovered in Campo Laborde, Argentina (about halfway between Buenos Aires and Bahía Blanca), where the rib bone of a *Megatherium americanum* appears to have butchering marks.

Putting aside human predation, the epoch after the Pleistocene, called the Holocene, marked the end of the earth's last major glacial period, and so the natural warming that came with climate change may have had a devastating impact on giant sloths, which had developed larger bodies to resist cold. Nor were they the only group to die out at this time; the great era of megafauna was soon to be in the past.

Let us return for a moment to the more detailed, yet important, features of Xenarthran taxonomy in an effort to understand this very primitive group and derive some insights about their descendants. Within the superorder Xenarthra is the order called Pilosa (the 'hairy' ones), which comprises the anteaters including their small, semi-arboreal relatives the tamanduas (a smaller and more lithe version of an anteater) and, more importantly, the sloths themselves, who belong to the suborder Folivora (or leaf-eaters). Armadillos, which are distinctly non-hairy, belong to the order Cingulata (which refers to being 'girdled' or 'banded'). Many texts still associate the outdated order 'Edentata' (toothless creatures) with sloths, but that grouping, which included the Old-World pangolins and aardvarks, was inaccurate. The Xenarthra is the modern superorder that was created to properly describe this group of exclusively New World species.

Although the extinct members of the Folivores are, as we have already noted, very impressive – at least in size, strength and even distribution – I will first consider the living members of the sloth family, and then look back at their phylogenetic connections. There are currently two genera of sloths: *Bradypus*, the three-toed sloths, and *Choloepus*, the two-toed sloths. Each is distinguished by the number of digits on their front limbs, although all species of sloth have three toes on their hind legs. It has become a growing trend, by the way, to refer to these sloths using the terms 'two-fingered' and 'three-fingered', given that the forelimbs of both animals resemble and often function like arms in primates. While the logic makes some sense (as did the once-common term 'Quadrumana' for most primates), I will use the older terminology for the duration of the book, or simply refer to the two groups as *Bradypus* and *Choloepus*.

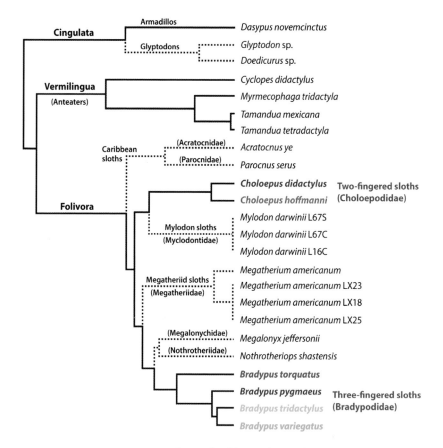

Cingulata
Armadillos — *Dasypus novemcinctus*
Glyptodons ┄┄ *Glyptodon* sp.
┄┄ *Doedicurus* sp.

Vermilingua
(Anteaters)
Cyclopes didactylus
Myrmecophaga tridactyla
Tamandua mexicana
Tamandua tetradactyla

Folivora
Caribbean sloths ┄ (Acratocnidae) ┄ *Acratocnus ye*
(Parocnidae) ┄ *Parocnus serus*

Choloepus didactylus   **Two-fingered sloths**
Choloepus hoffmanni   **(Choloepodidae)**

Mylodon sloths
(Myclodontidae)
Mylodon darwinii L67S
Mylodon darwinii L67C
Mylodon darwinii L16C

Megatheriid sloths
(Megatheriidae)
Megatherium americanum
Megatherium americanum LX23
Megatherium americanum LX18
Megatherium americanum LX25

(Megalonychidae) ┄ *Megalonyx jeffersonii*
(Nothrotheriidae) ┄ *Nothrotheriops shastensis*

Bradypus torquatus
Bradypus pygmaeus   **Three-fingered sloths**
Bradypus tridactylus   **(Bradypodidae)**
Bradypus variegatus

Evolutionary chart outlining the development of sloths and their Xenarthran relatives (anteaters and armadillos). Dashed lines indicate extinct lineages or species.

As an aside, we should note that sloths are not the only arboreal creatures to have lost their thumbs over evolutionary history. Spider monkeys, which are distributed from southern Mexico to Brazil, belong to the genus *Ateles* (a word meaning 'incomplete' in Greek), referring to their reduced or non-existent thumbs. The African Colobus monkey, which is noted for its 'stump-like' thumbs, derives its unflattering name from the Greek word *kolobós*, which means 'maimed'. The long and slender hands of gibbons appear to have reduced thumbs, and while the thumbs are not prominent, they are functional. Gibbons, of course, belong to the apes and can be found in the tropical forests of the Indian subcontinent, as well as in Borneo, Sumatra and Java. The bottom line here is that all of these animals brachiate – that is, move from limb to limb using their arms – through their respective tree canopies. Given the differences between these groups

George Edwards,
The Sloth, 1758,
colour woodcut.

– sloths, monkeys and apes – what we are observing is yet another example of convergent evolution among arboreal animals.

To digress a little further, at least geographically, it is worth noting the existence of tree kangaroos of the genus *Dendrolagus* in Australia. There are some similarities between tree kangaroos and sloths, particularly with respect to diet and digestion, but not only are these two radically different groups but the arboreal kangaroos, in contrast to arboreal sloths, were the precursors to the more familiar terrestrial kangaroos we know from Australian natural history. What's more, the development of large and more terrestrial species did not result in the extinction of the line emerging from their 'treed' ancestors.

The two families of sloths, though they share evolutionary origins, sit on opposite ends of the family tree and so their physical and behavioural resemblance, however uncanny it may seem, is the product of convergent evolution. In other words, both groups found themselves dealing with reasonably similar environmental constraints, which, as a consequence, resulted in two sets of

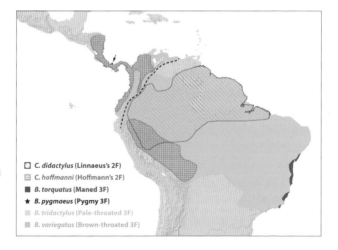

Map of the distribution of extant two-fingered (2F) and three-fingered (3F) sloth species across Central and South America.

☐ *C. didactylus* (Linnaeus's 2F)
⊞ *C. hoffmanni* (Hoffmann's 2F)
■ *B. torquatus* (Maned 3F)
★ *B. pygmaeus* (Pygmy 3F)
▨ *B. tridactylus* (Pale-throated 3F)
■ *B. variegatus* (Brown-throated 3F)

Michael Maslin's amusing, though erroneous, Bradypodid cartoon.

"*I thought you should hear this from me before you heard it from anyone else, Irene. I'm an arboreal tropical American edentate of the family Bradypodidae. In other words, a two-toed sloth.*"

animals that we perceive, at least superficially, to be alike. The confusion is understandable and is very common, as is clear from this wonderful cartoon from the *New Yorker* by Michael Maslin, in which the sloth, claiming to be a Bradypodid, misidentifies itself as a two-toed sloth. Of course, there is always the possibility – in the world of talking cartoon animals – that this sloth (perhaps a lothario among sloths, or to quote a colleague, a 'slothario') is playing a trick on the unsuspecting Irene, his female companion in the cartoon. But that secret rests with the sloth and, of course, Maslin.

However alike the sloth families may seem, they are substantially different, and so in order to fully understand all of the sloths, we need to be able to distinguish one from the other, beyond simply counting toes (or fingers) on the forelimbs. At the outset, we will begin with external appearances, starting with Bradypodidae, the three-toed sloths.

There are four species of three-toed sloths, or sloths under the genus *Bradypus*, which translates from the Greek to mean 'slow' (*brady*) 'foot' (*pous*): *B. variegatus*, the brown-throated sloth; *B. tridactylus*, the pale-throated sloth; *B. torquatus*, the maned sloth; and *B. pygmaeus*, the pygmy sloth. *Bradypus* forelimbs include three central 'fingers' that evolved from a five-fingered precursor. Each of those fingers has a lengthy nail or claw that is formed from an extensive and robust keratin sheath – that is, a nail – which is attached to strong phalanges, or finger bones. These sloths have very small external ears and a short tail of no more than 6–7 centimetres (2–2¾ in.), and their eyes have a slightly reddish hue.

Within this group, the brown-throated sloth, distinguished by its familiar dark-brown masking around its eyes and seeming smile, is probably the most frequently represented sloth in popular culture. It is also the most widely distributed species, ranging from Honduras to northern Argentina and covering areas in Colombia, Ecuador, Bolivia and the eastern forests of Peru. *Bradypus* sloths have blunt, dark noses compared to the more elongated and pinkish nose of *Choloepus*, and their colouration is typically darker. Sloth fur is at its darkest in what is called the corona – some might say a mullet – of deep-brown hair in the maned sloth.

Most *Bradypus* sloths are sexually dimorphic – at least at maturity, when a distinctly coloured dorsal patch, which is called the speculum, appears on the backs of males. These patterns (particularly pronounced in *B. variegatus* and absent in *B. torquatus*), which are unique to each individual, include bright patches of yellow, orange and white fur, typically surrounding a dark line

Two noses:
Choloepus (*top*)
and Bradypodid
(*bottom*).

that runs up and down the spine. What makes this feature odd is that sloths, which are generally camouflaged by green algae and by limited motion, would hardly seem to benefit from having bright markings. The evolutionary advantage of the markings is not clear, but they certainly play an important role in sexual selection. It may also be true that however conspicuous the markings seem to us, they might also function as disruptive colouration by breaking up the form of the sloth. If that is the case, there may be a balance between attracting mates who recognize the coloured patterns and avoiding predators who may be confused by them.

The maned sloth (*Bradypus torquatus*), which has been designated an endangered species by the International Union for Conservation of Nature (IUCN), can now only be found in the

Distinctive speculum on a male Bradypodid.

Maned sloth (*Bradypus torquatus*) in the Atlantic Forest, Brazil.

coastal rainforests of Brazil. The characteristic feature of the species is an impressive dark mane of fur, typically more pronounced in males than females, that surrounds the head and may appear on the chest. Females, however, are often larger (as much as 9.5 kilograms or 22 lb) than males (7.5 kilograms or 17 lb), which is true for sloths in general. The facial appearance of the maned sloth, which lacks the familiar eye-masks that characterize brown-throated sloths, is distinctive in two ways at least: first, the eyes are small and set narrowly apart; and second, the facial fur is a single colour, typically from light to dark brown. There appears to be a substantially different genetic profile for *B. torquatus*, and it is assigned, by some taxonomists, to its own genus, *Scaeopus*. Taxonomic designations aside, the evidence suggests that the

Early depictions of a maned sloth (Bradypodidae) and Linnaeus' two-toed sloth by Jean-Gabriel Prêtre, 1816–29.

Sloths with the 'sloth bear' (*Melursus ursinus*) from the Indian subcontinent, which is unrelated to the Xenarthrans. From John Wilkes, *Encyclopaedia Londinensis* (1796–1829).

maned sloth split off from other Bradypodids over 12 million years ago. The maned sloth is, at least by weight, the largest of all the three-toed sloths, yet, as far as we know, its lifespan may be a bit shorter. It typically eats leaves from *Inga edulis* (joaquiniquil) and *Mandevilla* vines as well as trees such as *Ficus* spp. and *Microphilus* spp., and of course the same *Cecropia* leaves that are eaten by all Bradypodids. In general, sloths tend to prefer younger leaves, part of new growth, because they are easier to digest. And as devoted as they appear to be to the *Cecropia* tree, they forage on a broad spectrum of leafy trees in the forest. In this way they avoid building up too many toxins from any one species.

B. tridactylus, or pale-throated sloths, often live up to their name by having a slightly yellowish throat patch, although from a distance, most are mistaken for the brown-throated sloth. Pale-throats, however, have a much more limited range than brown-throats, and are restricted to the more northern regions of South America, including Guyana, Suriname, French Guiana and limited parts of Venezuela and Brazil.

As with many sloths, there is a microcosm of insects in the coat of *B. tridactylus*, including the sloth moth, which lives in the fur and whose larvae, hatched from eggs in sloth manure, feed on dung. Other insects, variants of the scarab beetles, *Trichillum adisi* and *Uroxys batesi*, can inhabit the fur of sloths as well as on dung pellets, which may also support their larvae. Sloths, because of their slowness, also attract ticks, mites and mosquitos, but it doesn't seem that these insects (and arachnids) have serious negative effects on the animals in spite of the fact that sloths do not groom each other as so many primates do.

The pygmy sloth (*Bradypus pygmaeus*), which was only designated as a distinct species in 2001, is limited to the tiny island of Isla Escudo de Veraguas, 17 kilometres (10.5 mi.) from the Panamanian mainland. These sloths, sometimes called dwarf sloths or monk sloths, which measure about 50 centimetres (20 in.), and

Fanciful picture of yellow-throated sloths featuring eerily human faces.

weigh around 3 kilograms (6 lb), are considerably smaller (though not always shorter) than the brown-throats on the mainland. They are the most endangered of all the sloths, with numbers that hover around eighty individuals. Their reduced body size is surely an adaptation, called insular dwarfism, that stems from having been isolated – perhaps originally as brown-throated sloths – in a very small island region with limited resources for close to 10,000 years. A similar phenomenon occurred with both humans and elephants on the island of Flores in the eastern archipelago of Indonesia. 'Flores man', *Homo floriensis*, was probably no more than 1.1 metres or 4 feet tall, weighing around 25 kilograms or 60 pounds. The island also sustained an elephant, *Stegodon*, that was a little more than 2 metres (6 ft 6 in.) at the shoulder and that weighed around 400 kilograms (800 lb).

The pygmy sloth can be found in red mangroves, the leaves of which are even lower in nutrients than the *Cecropia* leaves eaten by brown-throats. Perhaps because of the low level of nutrient intake, these sloths seem to be the slowest of the entire Bradypodid family. The pygmy sloth has a smaller and more slender skull than its relatives, but researchers have noted that it has a larger ear canal. Whether this means greater hearing sensitivity than

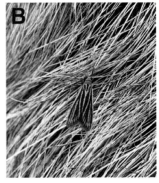

Sloth moths on the nose of a Bradypodid (A) and a close-up of the moth itself (B).

other Bradypods is not clear. But given that these sloths feed on red mangroves, which are no more than 6 metres (20 ft) tall, the sloths are more vulnerable and so there may be a considerable advantage in a more acute sense of hearing.

CHOLOEPODIDAE: THE TWO-TOED SLOTH

The two-toed sloths belong to the genus *Choloepus*, which is Greek for 'lame-foot' and harkens back to a period when the ungainly nature of sloths when placed on the ground was a source of derision. Two-toed sloths are, of course, anything but 'lame' in their natural habitat, where they are comfortable climbing trees, searching for food and securing rest for themselves in the nooks of trees. The two toes are, once again, two forefingers, as opposed to the three of the *Bradypus*, but like their distant cousins, they have three toes on their hind legs. *Choloepus* ancestry can be traced back to a root member of the Mylodont family, but the two species of *Choloepus* are themselves ancient and probably diverged from each other over 6 million years ago. The earlier ancestors of the

A typical two-toed sloth.

Choloepus not only included the Mylodonts, but other large ground sloths also found in both South and Central America as recently as 10,000 years ago. Their ancestral ground sloths, the Megalocnidae, could be found in select areas – Cuba, Hispanolia, Puerto Rico and some other islands in the Greater and Lesser Antilles in the Caribbean. Some species, such as those in the genus *Megalocnus*, weighed as much as 400 kilograms (nearly 900 lb) and were unquestionably terrestrial, while other species, such as those in the genus *Neocnus*, which may have weighed as little as 10 kilograms (22 lb), were surely arboreal.

For all the similarities of the two groups of sloth, the evidence suggests that they are 'diphyletic'. In other words, whatever shared origins they may have had so far back (perhaps 40 million years), they simply can't be described as descendants of the same or even a similar group. Thus, similar traits, including hanging upside down, occurred in parallel. Biologist Timothy Gaudin, who has written extensively about sloth origins, is himself nonplussed by the incredible correspondences between *Bradypus* and *Choloepus*. 'Similarities between the two taxa, including their suspensory locomotor habits,' he writes, 'present one of the most dramatic examples of convergent evolution known among mammals.'[2]

As with most examples of convergent evolution, both genera of sloth evolved to adapt favourably to the environmental pressures of their habitats. And yet, one can't help but be amazed that two groups of giant ground sloths somehow found similar evolutionary paths to becoming what we now generically call tree sloths.

A close-up of the two digits and claws on a two-toed sloth.

An interesting footnote to the emergence of similarities in the tree sloths is the now-extinct Jamaican monkey (*Xenothrix mcgregori*), which disappeared roughly nine hundred years ago. Evidence based on the arrangement of the monkey's teeth and the structure of its thigh bone suggests that the species – isolated on Jamaica – may have been evolving sloth-like traits. What might have come to pass had they survived can never be known, but this does remind us that convergence, though it requires many millennia, is an important response to environmental constraints in the course of evolution.

Because of their substantial convergence, which often masks important differences in the two groups, a detailed look at the behavioural and physical characteristics of two-toed sloths and three-toed sloths makes good sense. First, let us turn to the two-toed sloths, traditionally called 'unau', a name drawn from the language of the Tupi, who were among the most numerous Indigenous populations that inhabited what is now Brazil. Typically larger than the three-toed sloths, two-toed sloths often have straw-coloured fur and a more prominent and pinkish snout. *Choloepus* noses frequently appear to be moist, and for good reason: it is the only place where *Choloepus* are capable of sweating, and perspiration is important in regulating their temperature.

What's more, two-toed sloths have no tail, an appendage used by Bradypods along with their hind legs to dig small holes for their faeces. While they don't have an undercoat as three-toed sloths do, their fur is often long and also harbours algae, which lends the animals a greenish hue. The diet of two-toed sloths is also more diverse than that of the Bradypodidae. In addition to leaves, two-toed sloths will eat small insects, tree buds and bark, fruits and berries, and often flowers, particularly hibiscus petals. Their more varied diet makes these sloths more suitable for captivity, although the survival rate in captivity is still not impressive, nor is it commendable. While *Choloepus* have canine-like teeth and will use them to bite predators, their teeth are blunt and are essentially used for grinding plant materials. If faced with the option, they tend to prefer younger leaves, which are easier to break up and digest. All of these materials are processed through a complex four-chambered stomach and is only excreted once every four to five days or more. Storing urine and faeces for a prolonged period can give some sloths the appearance of being bloated, and X-ray images have, in the past, been taken of captive sloths to determine whether the bloated sloths were in good health. As it turns out, after relieving themselves, they were clearly vigorous and sound.

Although the Tupi word *unau* seems to refer to both species of two-toed sloths, they do have separate names in Western classification: *Choloepus didactylus*, sometimes called Linnaeus' two-toed sloth, and *Choloepus hoffmanni*, or Hoffmann's two-toed sloth. The former is, of course, named for the great Swedish taxonomist Carl Linnaeus (1707–1778), who probably never saw a living sloth, while the latter is named after the German naturalist Karl Hoffmann (1823–1859), who took up residence in San José, Costa Rica, where he was also a pharmacist and doctor. Linnaeus' sloth can be found in Venezuela, as well as in the 'Guyanas', which includes Guyana, Suriname and French Guiana, and in Colombia, Ecuador, Peru and the upper latitudes of Brazil. The range of Hoffmann's sloth overlaps, to some extent, with Linnaeus' and is present in Costa Rica and Central America; it ranges more to the western side of the continent and can be found in Ecuador, on the eastern side of the Andes.

Both species have long fur that is parted in the middle of the abdomen to allow for better water flow while suspended upside down. They share similar colouration, but Hoffmann's sloth has a darker, almost purplish, snout, as well as darker markings on the forelimbs and the legs. Both have pronounced eyes surrounded by a small dark ring of fur. Aside from minor differences in their coats, the two species are difficult to tell apart from a distance. The arms of both *Choloepus* species are longer than those of the Bradypodidae, which perhaps makes sense, as they are more mobile and have broader tastes. But the fact is that the arms of both groups tend to be twice as long as their legs. The teeth, as with three-toed sloths, lack enamel and continue to grow through the animal's lifetime. The two front teeth in *Choloepus* have the appearance of canines and although technically they are not, they can still inflict a nasty bite. *Choloepus* are equipped with an extraordinary number of relatively flexible ribs that may be useful in helping them avoid injury during a fall. Both species of *Choloepus* are nocturnal and spend a lot of time suspended upside down from trees,

Two-toed sloth resting in the crux of a tree.

The ocelot, a primary predator of sloths.

A captive tayra (*Eira barbara*), a member of the weasel family that preys on sloths.

although they will seek refuge in tree crevices when not active. They must, like their *Bradypus* relatives, descend the tree that they are in so as to urinate and defecate, and when they do, they are vulnerable to predation from ocelots, jaguars, margays, anaconda and tayra (a kind of weasel). Their main predatorial concern in the treetops are harpy eagles, which are powerful predators that can weigh anywhere from 6 to 9 kilograms (13–20 lb).

Choloepus have been known to live for close to forty years in captivity, although their lifespan in the wild is certainly much

Harpy eagle in flight revealing the powerful talons, lethal to many sloths.

Harpy eagle in flight revealing the powerful talons, lethal to many sloths.

shorter. A sloth that lives to seventeen years, an age that has been recorded for a tagged *Choloepus* in the wild, is an old sloth indeed. Courtship is initiated by the call of the female, which can attract several males who are nearby. After a male has been selected and accepted, copulation is not an extended process, but it is worth noting that in addition to conventional postures, sloths may mate face to face. Once a female is pregnant, gestation will last around six months for the Linnaean sloth or close to a year for Hoffmann's, and males have long disappeared well before there are any signs of pregnancy, in fact almost immediately after copulation. The gestation period for Bradypods is much shorter, and pregnancy rarely lasts longer than about six months.

While the reproductive system of sloths varies a bit from the typical mammalian model, the pattern is generally that of a single offspring born vaginally. Giving birth in the forest canopy, however, can make for a difficult balancing act for both mother and infant. We must assume that sloths take care to cradle themselves – that is, usually upside down – in such a way as to be able to retrieve their newborns and to carry them up to their chests. On occasion, an infant might dangle precariously from the umbilical cord, only to be successfully retrieved. Nevertheless, more than a few newborns and sometimes even juveniles have been known to tumble to the forest floor. The prevailing myth is that once offspring fall to the ground, they are abandoned by the mothers, yet there are many cases in which mothers have descended (albeit

Nineteenth-century engraving of a harpy eagle, which is supposedly able 'to split a man's skull with a single blow of its beak'.

slowly) to retrieve their young. Fallen infants do emit pitiful cries to draw their parent's attention, but in some cases, those cries attract the notice of unwitting passers-by, who ostensibly 'rescue' the young.

Infants are precocious when born and have fur and claws that allow them to hold tightly onto their mothers. They take solid food fairly quickly – after two weeks or so – but do nurse for up three months. Producing milk demands a good deal of energy

Infants holding on to their mothers.

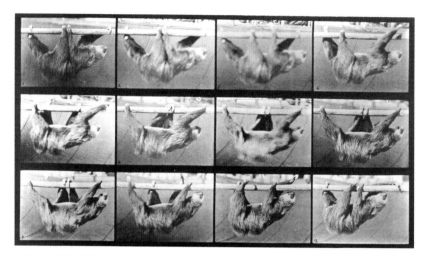

and considerable nutrients, so it isn't surprising how short the nursing stage is in both *Bradypus* and *Choloepus*. Nor is much milk produced or stored at any one time. The teats of the sloth, in all species, are located on the upper portion of the chest, but most sloths are usually observed (if observed at all) nursing only very young infants.

It is very difficult to monitor sloths in the wild, particularly when their behaviours are nocturnal, and thus much of what we know is drawn from serendipitous observations in the wild or from behaviour observed in captivity, particularly in the sloth rescue shelters that are dotted through Costa Rica and Colombia. Fascinated by the upside-down gait of two-toed sloths, Eadweard Muybridge (1830–1904), the great photographer of animals and people in motion, included a series of photographs titled 'Sloth Walking Suspended on a Pole' in his book *Animal Locomotion* (1887). The images reveal a left–right pattern in which one set of limbs is extended while the other set is close together. Whether this always holds true when walking suspended from trees or

Eadweard Muybridge's study of sloths in motion, c. 1884–7, collotype.

43

perhaps climbing them is not always clear, but the pattern does appear to allow a steady and secure gait. They may be more visible than we think while ambling through the canopy at night. It has been noted by Kenny J. Travouillon of the Western Australian Museum that *Choloepus* fluoresce under ultra-violet (UV) light, a phenomenon that he has documented in other animals.[3] Whether fluorescing provides any benefit to the sloth remains uncertain, but it may offer some advantages in these primarily nocturnal animals. Future research will surely reveal whether sloths can actually detect UV light and thus see each other in the dark. Equally interesting is whether predators might be able to take advantage of fluorescing when hunting sloths.

Choloepus appear to fluoresce under ultra-violet light – a recent discovery.

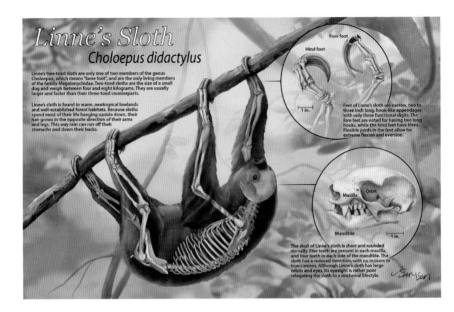

Linne's Sloth
Choloepus didactylus

Linne's two-toed sloth are only one of two members of the genus Choloepus, which means "lame foot", and are the only living members of the family Megalonychidae. Two-toed sloths are the size of a small dog and weigh between four and eight kilograms. They are usually larger and faster than their three-toed counterparts.

Linne's sloth is found in warm, neotropical lowlands and well-established forest habitats. Because sloths spend most of their life hanging upside down, their hair grows in the opposite direction of their arms and legs. This way rain can run off their stomachs and down their backs.

Fore foot

Hind foot

1 in.

Feet of Linne's sloth are narrow, two to three inch long, hook-like appendages with only three functional digits. The fore feet are noted for having two long hooks, while the hind feet have three. Flexible joints in the feet allow for extreme flexion and eversion.

Orbit

Maxilla

Mandible

1 in.

The skull of Linne's sloth is short and rounded dorsally. Five teeth are present in each maxilla, and four teeth in each side of the mandible. The sloth has a reduced dentition, with no incisors or true canines. Although Linne's sloth has large orbits and eyes, its eyesight is rather poor relegating the sloth to a nocturnal lifestyle.

2 Anatomy and Physiology

One more defect and they could not have existed.
Georges-Louis Leclerc, Comte de Buffon (1707–1788), on sloths[1]

The question as to why sloths are so slow is not as simple as one might think. The problem is one of what is called 'adaptive fitness', a principle which suggests that – in general – behavioural and/or physical modifications should ultimately benefit an animal. Slowness, at face value, may well seem maladaptive because it makes sloths more subject to predation, since they cannot escape predators with any speed. But the alternative, and this is surely one of the keys to their survival, is that they can be so slow and stealthy as to go unnoticed by predators in the undergrowth and the tree canopy. Slowness is thus an adaptation of some kind, but to what? The clues that we have from palaeontology are not that helpful, but what is certain is that sloths, apparently vegetarians (predominantly) from their origins, gradually limited themselves to a very narrow diet and hence a very restricted ecological niche.

Sloths are primarily folivores, which means that the primary food source for sloths is leaves, most frequently from *Cecropia* trees, a setting which also provides ample shelter for the foraging animals. It probably goes without saying that a diet of leaves alone does not provide a lot of nutrition, a characteristic exacerbated by the fact that sloths must masticate and process the leaves to extract all of the nutrients possible. With low levels of protein and very long periods of digestion, it makes sense to stay still and not expend too much energy.

The skeletal structure of a two-toed sloth.

Two-toed sloths eat primarily the same kind of diet as their three-toed counterparts, although they are more likely to add some insects, fruit and hibiscus petals (which they seem to delight in) to their diet. When all is said and done, this diet, which offers only slightly more protein than plants alone, is well suited for a quiet and relatively inactive existence, particularly because it also requires a lot of chewing. The sloths use blunt, flat, grinding and enamel-less teeth to wear leaf fibres down, and so, quite appropriately, the teeth continue to grow throughout the lifetime of the sloth. Once worn down in the mouth, the leafy materials enter an even slower process of digestion in the gut. To break down the cellulose, the digestive system – which includes a four-chambered stomach – is equipped with microorganisms in the first three chambers that break the cellulose into basic components for the more conventional final stomach.

Compared to other herbivores, even to other creatures living in the tree canopy, the sloth seems to eat relatively little per day. Because of the slowness of the digestive process, their stomachs

Leaves of the *Crecropia*, a primary food source for sloths.

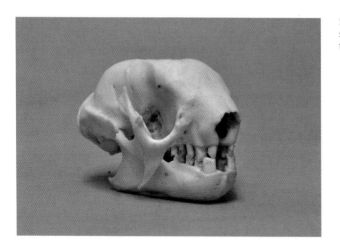

Skull of a three-toed sloth showing the teeth.

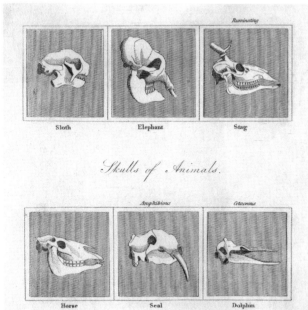

The sloth skull in context, from *The Atlas of Nature* (1823).

remain fuller longer and so, it seems, they are seldom driven by extreme hunger. *Choloepus* are capable of chewing their cud, as it were, by returning leaves to the mouth for additional mastication, and it has been speculated that on sunny days, sloths lie with their stomachs exposed to the sun in order to expedite fermentation.[2] Satiation is partly enabled by the 'forestomach', where most of the very slow fermentation takes place for a period lasting up to 150 hours. During the entire process of digestion, a good deal of water is extracted from the leaves, thus precluding the need to descend from the forest canopy to drink.

One of the great advantages of remaining full and digesting slowly is that the sloth defecates and urinates much more infrequently than other animals. Typically, sloths relieve themselves once a week, which is another adaptive advantage given how vulnerable to predators they are when on solid ground. When they feel the urge, they descend from their perch in the canopy and make their way down to the ground, where they leave very distinctive-looking scat consisting of hard pellets. The act of defecating has the appearance of being very satisfying, but no doubt that's simply – or at least only partly –an anthropomorphic response on our part as we imagine the relief entailed in a weekly 'bathroom' visit. That relief, if that is what they are experiencing, is surely counterbalanced by fear or anxiety. The very fact that they are on the ground, where their mobility is limited and where they may be more conspicuous, is hardly advantageous for a sloth. The desire to climb back up to safety must be at least as urgent as the pressing necessity of relieving itself. Still, before ascending, three-toed sloths will often use their tails to dig a little hole for their excrement, which, of course, is home to the eggs and larvae of the sloth moth.

Why sloths have evolved to defecate on the ground, as opposed to simply dropping their faeces and urine willy-nilly (if

Example of sloth faeces, which are pellet-like.

you can forgive the phrase), as most arboreal animals do, remains an open question. Two theories have been proposed for this idiosyncratic behaviour. One is that the animals are scent-marking trees (although the animals themselves can be itinerant), or that the behaviour allows sloth moths to easily deposit their eggs in the faeces. The moths seem to 'cultivate' the algae in the sloth's fur as well as introduce more nutrients (such as nitrogen), either by importing small amounts of faecal matter or when some actually die while on the sloth. Another, more speculative, theory about why sloths descend trees in order to relieve themselves has been advanced by Bryson Voirin, who argues that in the process of climbing down to the ground, sloths manage to scrape up dirt filled with minerals missing in their daily diet that they later ingest.[3]

When we understand the food intake and the protracted digestive process of sloths, other aspects of the sloth's rather complex physiology and anatomy make more sense. Although clearly strong climbers and well equipped to hang from and to move across the tree canopy, sloths have very little muscle mass, which has been reduced in part (quite probably) because of low levels of dietary protein. Lower levels of sugars and carbohydrates, the fuel for most mammals, also result in slow and more deliberate movements. Like all other mammals, the sloth is warm-blooded, but it typically maintains a very low body temperature ranging from 30° to 34°C (86° to 93°F) and has a metabolic rate that is often 50 per cent lower than comparably sized mammals.

Given the levels of reduced activity in the sloth, it isn't surprising that it has one of the slowest metabolisms of all mammals. With little activity to elevate its temperature, the animal stays at rest for as long as possible. If the typical mammalian temperature hovers about 36°C (97°F), which is ideal for an active and dynamic life in most climates, the sloth typically averages around 27°C, or 80°F, which means that the animal is burning less energy and thus is less vigorous. To be sure, this temperature is still higher than in most reptiles but clearly less than that of birds. As a consequence, the sloth moves slowly and methodically and thus conserves the limited energy in its system. Three-toed sloths have a lower metabolic rate on average than two-toed sloths, perhaps because their more limited diet requires less overall movement.

The slowness of sloths is thus not simply behavioural, but a by-product of a system adjusted to a diet low in nutrients and a structure adapted to a very restrained arboreal life. Their physiology of movement is such that the contraction rate of muscles is anywhere from four to six times slower than for animals of comparable size (the cat, for example). As a result, sloths are simply incapable of fast responses, and the value of a 'fight or flight' response is almost non-existent. That's not to say that sloths are incapable of inflicting injuries on predators or intruders with their sharp claws. And two-toed sloths can be very effective biters, often leaving serious and sometimes festering wounds on their opponents.

Sloths may not need highly reactive muscles, but their tendency to suspend themselves, in various positions, from tree limbs requires elaborate retractor muscles. When these muscles are contracted, a sloth can hold tight to a tree while reaching out for leaves or fruit, or, as is more common, suspend themselves upside down.

However simple it seems to the observer, living the bulk of one's life upside down is very demanding with respect to anatomy and physiology. The gravitational effect of this behaviour has also led to interesting and even unique mammalian adaptations. It is well worth considering the flow of blood in an animal suspended

The strength of sloths, either to extend or to suspend themselves, is very impressive.

upside down. Sloth hearts are neither large nor powerful, affecting both blood pressure and pulse, the latter of which never exceed 110 beats per minute. To ensure the proper flow of oxygenated blood to the limbs, the circulatory system relies on a complex web of minute arteries, a 'rete mirabile' (wondrous net), that are entwined with similarly complex veins. The minute arteries help keep oxygenated blood flowing in limbs that are often not in motion for extended periods of time. This arrangement may help prevent deep-vein thrombosis (DVT), a condition in humans that is often exacerbated by lengthy periods of inactivity. The proximity of the veins and the arteries in the rete may also help sloths maintain their body temperature by offloading or conserving blood warmth as necessary. The advantage of retaining warmth, particularly during cool evenings, is a decided advantage for an animal that is unable to shiver.

The act of eating is one of the more energetic periods in the life of the sloth, and thus it is not surprising that both the blood pressure and the rate of breathing increase at 'meal' times. While humans typically consume 5 per cent of oxygen breathed in, sloths function at only 2 per cent. Even more striking is the fact that sloths can recover after having their breathing stopped for periods up to twenty minutes, but of course this phenomenon involves a dramatic reduction of their metabolic rate. Whether sloths would ever 'need' to stop breathing outside of laboratory settings is a question worth asking. Still, the ability to reduce breathing and metabolism can be useful when eluding predators who are sensitive to even the slightest motion. A similar behavioural phenomenon, called 'alarm bradycardia', has been observed in fawns of white-tailed deer, who, when anxious due to the threat of a predator, become motionless and reduce their breathing considerably.[4] The bradycardia strategy (invoking the same root word in Greek for slowness – *brady* – that identifies *Brady*podia) is probably instinctive and helps reduce detection by predators; thus, it is as useful for the deer population as it is essential for sloths.

ANATOMY

Sloths enjoy an advantage, in their hands and feet, of having tendons that automatically allow the sloth to grasp and hold onto tree branches without having to exert any substantial energy. In this they resemble birds, another tree-dwelling group, which have toes that will grasp a perch automatically when they land. In other words, neither sloths nor birds need to grasp a surface deliberately or actively in order to hold on to it, which is a highly advantageous system for sleeping, resting and simply remaining in place for an extended period.

Spending so much time in what is typically called an inverted state requires adaptations of the internal organs of the sloth. One can be forgiven for not considering the fact that while upside down, many conventional organs in the sloth press down against the lungs, and the diaphragm thereby reduces the efficiency of breathing. And as the extensive stomach system continues to fill with slowly digesting leaves, this pressure might be distressing indeed. Recent findings in the variegated sloth indicated that sloth organs are actually suspended by fibrinous adhesions on the lower rib cage to prevent them from compressing the lungs, thus reducing the energy required for breathing.[5]

While virtually every mammal, from mice to humans, dolphins and giraffes, has seven cervical (neck) vertebrae, sloths typically have more. The three-toed sloth has nine, while the two-toed can have anywhere from six to nine. The additional cervical vertebrae, which are essentially borrowed from lower (thoracic) spinal vertebrae, presumably allow the sloth to rotate its head more

Sloths are capable of rotating their heads up to 270 degrees.

comprehensively, which benefits the animal while feeding and perhaps while keeping an eye out for predators from both above and below. The three-toed sloth is able to turn its head a remarkable 270 degrees, and thus sloth-watchers often note the uncanny ability of the animal to look directly at you – right-side-up – from almost any position.

No less uncanny, but obviously quite sensible, is the pattern of growth of the sloth's fur, which, as I indicated earlier, grows out and downward from the belly or abdomen in two-toed sloths and in a similar pattern – from the hindquarters – in the three-toed sloth. This surely offers an advantage with respect to water running off the body in the many rainstorms experienced by the animals. Sloth fur is coarse and, in some cases, grooved, which provides ample surface for algae to grow on the animal's coat. The greenish tint provided by the algae, in the wet season, is presumably helpful in camouflaging the slow-moving sloth. This mode of camouflage was apparently widely known, or so thought the London and Manchester Assurance Company (L&M), which included a sloth in an early advertising campaign with the tag line, 'The Three Toed Sloth is protected by the camouflage of minute plants which grow on its fur. Let your protection be L&M.' At the very least, it was an interesting gambit to tout the virtues of insurance.

The sloth's fur, as depicted in this illustration by Sylvaine Peyrols, is adapted to an inverted posture and directs rain downwards.

The striking green colouration of the animals, whose fur ranges from grey to brown – sometimes covered with lichens or arboreal debris – helps disguise the sloths at all times. Sloth fur is also home to a number of invertebrates, such as beetles and ticks, which apparently find refuge tucked away in the undercoat of the animal. One insect in particular, the snout moth or 'grass moth' (Pyralidae), has developed a beneficial relationship with the sloth. The moth, which inhabits the fur of sloths, particularly *Bradypus*, lays its eggs in the faeces that the sloths deposit every week or so on the ground. The larvae consume the excrement of the sloth and, after they mature, each flies up to the canopy to find its own new sloth host. While the exact nature of the relationship is unclear, it appears that sloths that are populated with more moths tend to have more algae and nitrogen in their fur, which they can consume and consequently become more robust.

Scanning electron micrographs of sloth hairs in which algae can grow. *Bradypus variegatus* (A) has hair with transverse cracks that increase with age (from bottom up, young, mature, and old). *Choloepus hoffmanni* hair (E) is distinguished by longitudinal ribs.

Despite their furry coats, sloths are much lighter and smaller than they appear to be from a distance. Three-toed sloths typically weigh about 4 kilograms (9 lb), and two-toed sloths, which are often bigger, can weigh as much as 12.5 kilograms (27 lb). The pygmy sloth weighs in at no more than 3.5 kilograms (7½ lb) and is about 0.5 metres (1½ ft) long.

However awkward sloths are when they find themselves on the ground, they are reasonably good, if somewhat ungainly,

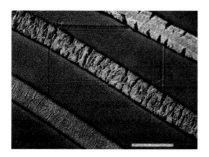

The Three Toed Sloth is protected by the camouflage of minute plants which grow on its fur

LET YOUR PROTECTION BE

LONDON and MANCHESTER

ASSURANCE COMPANY LIMITED

Finsbury Square, London, E.C.2.

Advertisement for the London and Manchester Assurance Company, 1960s.

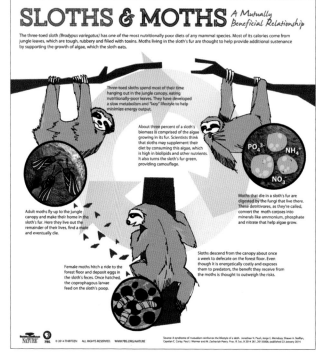

The life cycle of the sloth moth, designed by Karen Brazell.

Two-toed sloth swimming.

Three-toed sloth swimming comfortably.

swimmers. The low muscle mass seems to facilitate their buoyancy, and their strong retractor muscles enable a decent stroke so that they can advance. Faced with the challenges of small streams in the forest, which might block their path to food sources or, even more significantly, to potential mates calling from a distance, sloths often need to ford bodies of water to get around. It is worth noting that an entire group of prehistoric ground sloths, the Thalassocninae, were excellent swimmers and adapted to a partially aquatic lifestyle. All five species of the group, which lived from

3 to 6 million years ago, appear to have developed on the western shores of South America, in what is now Peru. Recent findings indicate that at least four species may have also inhabited the coastal waters and coastline of what is now Chile. While we cannot be certain why these animals adapted to a quasi-aquatic life, the more interior areas of Peru and Chile were very arid, and vegetation was increasingly limited, inducing the animals to seek a more supportive environment. Because of their remarkable bone density, they achieved a degree of negative buoyancy that facilitated grazing for sea grasses on the ocean floor. Their movements and activities along the shallow ocean floor were probably roughly similar or comparable to the way that hippos pad along riverbeds, but the sloths undoubtedly clawed their way forward. Seeking refuge in the water came at a cost, however, and the Thalassocninae probably had to develop some additional insulation below the skin to resist heat loss. And there were also threats from predators, most notably from what were called macroraptorial sperm whales of the genus *Acrophyseter*. Although only 3.9–4.3 metres (13–14 ft) long, the whales had an impressive array of teeth in the upper and lower jaws and were certainly powerful enough to prey upon the smaller and less agile aquatic sloths.[6]

Thalassocnus in its swimming posture, illustration by Carl Buell.

Swimming *Thalassocnus* under attack by Acrophyseter (ancient sperm whales), illustration by Alberto Gennari.

Recent investigations into the microbiota of sloth hair reveal a broad array of 'bioactive strains of fungi' that seem to resist the parasites that cause malaria and Chagas' disease, with some suggestion (using human cells) that they also have a negative effect on human breast cancer cells.[7] While mammals in general are distinguished by hair, sloth 'coats' are particularly complex and well worth further study.

VISION

The sensory world of the sloth, though no doubt rich for sloths, may seem rather limited compared to other animals. Recent studies have uncovered 'rod monochromacy' as part of the evolutionary development of sloths. What that means is that sloths, over time, have lost the cones that enable colour vision and so the animals can only see the world in black and white. Whether this is adaptive or not for two-toed sloths, the reality is that they are mainly nocturnal, and their vision is, in any event, occluded by the leaf cover in the forest canopy. They are able to see close up fairly well, which helps with grasping nearby branches, but given that they move slowly, the kind of visual precision that we associate with, say, birds is unnecessary. What is more, with a limited home range, and the ability to mark trees with their scent, they manage to get around quite well. Three-toed sloths also tend to be myopic but may be able to focus on objects a bit more sharply, although also in black and white. Among animals of both genera of sloths, bright light is certainly uncomfortable, if not, in some cases, blinding. Because both groups have eyes that face forward, there is certainly decent binocular vision, while peripheral vision is compensated for by the remarkable shape and rotational ability of their heads. Moving slowly, methodically and silently, the sloth – despite any visual deficiencies – is able to visually encompass the world around it in a manner unique to this particular mammal.

SOUND PRODUCTION, HEARING AND SMELL

Sloths are not vocal creatures. When in danger, they can produce a shrill whistle-like sound which, notwithstanding their solitary nature, may serve to alert nearby creatures of a threat. It is possible that the cry may be used to startle potential predators as well, but that seems unlikely given that such a call would also help identify the location of the animal. The sound produced by three-toed sloths, as represented in the Tupi language, is a distinctive 'ai-ai' (a-ee), which has come to be the common name of the sloth as well. What can only be called a shrill scream is emitted by females

who are ready to mate. These calls are repeated so that nearby males can find the females for what is ultimately a very brief moment of intimacy. Given the distribution of the animals, mating calls are necessary given that the nearest mate may be relatively far from the caller. Female sloths in oestrus (in other words, ready to mate) also make the dangerous journey to ground level in order to scent-mark their tree, which is yet another enticement to males. And the incentives, vocal and olfactory, are clearly effective, as females may mate with several males during their oestrus cycle.

The hearing range for most sloths ranges from 2 kHz to 8 kHz, which, from an acoustic perspective, is within the range of human

hearing and hardly exceptional in any way. When sloths are separated from their mothers, they produce a single rising tone for
several seconds, while a distress call among two-toed adults is
generally lower pitched and is often likened to a bleating sound.
Choloepus tend to hiss when sensing a threat, and given their ability to bite, the hiss is ample warning to potential predators. Since
sloth vocalizations are not sophisticated, it shouldn't be surprising that external ears are not prominent in sloths. They do exist,
if one looks carefully under their fur, but the fact that they are so
inconspicuous suggests that sloths do not rely on hearing for
survival.

Giant ground sloths, which are often depicted with more conspicuous ears or pinnae, may have produced sounds of a much
lower frequency, which would allow the vocalizations to travel
further and with less interference. Still, very little has been established about the vocalizations of prehistoric sloths, although
knowing as we do that they were certainly less hidden in the
landscape or in the tree canopy, their vocalizations may have
been radically different from those of their descendants.

But the ear is not merely an organ for hearing: we also know
that the inner ear of animals is important for balance and the
perception of movement. And it is striking that this mechanism
is not well developed in sloths. It is possible that the slower and
more measured movements of the sloth meant that there was less
need for a highly responsive sense of balance, and so little development (and possibly regression) has taken place over evolutionary
time.

If we are looking for precise sensory systems in sloths, they
may not be the kind of heightened senses that we find in other
animals. Brain studies of *Choloepus* indicate that the olfactory
bulbs are reasonably well developed for its brain size, perhaps
because of the diversity of their diet. But the sense of smell in the

sloth is believed to be far less sensitive than in its cousins, the anteater and the armadillo, which, as insectivores, need to be alert to the whereabouts of ants, termites and other invertebrate prey. Nevertheless, sloths do mark their territories with anal scent glands, which suggests that olfaction can play an important role in the distribution of animals and may also play a role in the bonding of mother and infant.

SLEEP

The standard and widely accepted perception of resting behaviour in sloths, based on studies in captivity, is that they are prodigious sleepers. Data recorded from captive sloths indicate that they can sleep for up to sixteen to twenty hours a day, which is a remarkable amount. More recent studies, albeit preliminary, of animals in the wild indicate that sloths may only sleep in the region of ten hours,

Captive *Choloepus* yawning and stretching out its tongue.

J. J. Grandville's cartoon depicting the sloth as the most prominent sleeper among his animal companions, 1842. The caption reads: 'Those who are not sleeping are yawning or about to yawn.'

and most of it, at least in three-toed sloths, is at night. This data reflects the reality of sloth life in that sleep is a perilous activity that puts the sleeper in danger of predation. It is only common sense that sloths need to retain a level of alertness about predators, of course, but also about interlopers – male or female – in their territory.

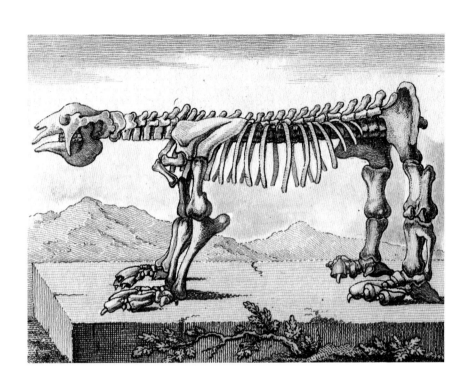

3 Sloth Species and Evolution

> I have been wonderfully lucky, with fossil bones. – some
> of the animals must have been of great dimensions:
> I am almost sure that many of them are quite new; this
> is always pleasant, but with the antediluvian animals it
> is doubly so. – I found parts of the curious osseous coat,
> which is attributed to the Megatherium; as the only
> specimens in Europe are at Madrid (originally in 1798
> from Buenos Ayres) this alone is enough to repay some
> wearisome minutes.
> Charles Darwin, letter to Caroline Darwin, 24 November 1832[1]

While we know where sloths come from, we are much less clear on how they actually came about. The palaeontological origins of the sloth can be best described, to borrow the words of a recent scholar, as 'an evolutionary conundrum'.[2] We have so little evidence of early sloths that some palaeontologists simply categorized ancestral sloths, rather awkwardly, as 'unallocated basal folivorans'. However we understand the term 'unallocated', the implication is that there is an enticing veil behind which we will discover a remarkable and exceptional evolutionary narrative. But that, as current researchers consistently remind us, will have to wait. Nevertheless, we can trace the earliest sloths to an era, following the Cretaceous period, called the Palaeogene, beginning roughly 66 million years ago. In that era, which is renowned for the emergence of diverse mammalian species, relatives of very primitive anteaters (Vermilingua) began to expand into new ecosystems and adapt to new environmental pressures. An early sloth form, the genus *Pseudoglyptodon*, which emerged about 44 million years ago (in the Eocene), can be considered an important precursor to the animals that we have come to know as sloths. But were those early sloths terrestrial or arboreal, or perhaps both?

Early tinted print of *Megatherium americanum* by Juan Bautista Bru, from Georges-Louis Leclerc, Comte de Buffon, *Histoire naturelle, générale et particuliere* (1799).

So far, we cannot answer that question, and the answer, or answers, may remain hidden forever.

It has been established, however, that the two major groups of sloths which we know, two-toed (*Choloepus*) and three-toed (*Bradypus*), developed separately from each other, and what we now observe as similarities owes little or nothing to relatedness; they are designated, in evolutionary terms, 'convergent forms'. In other words, their similarities were developed as independent responses to similar environmental constraints, not unlike the development of flukes in both dolphins and manatees, which are also unrelated. It is perhaps human inclination (or even arrogance) that, unfortunately, tends to group all sloths together, but despite their remarkable similarities, it should become clear that they are separate and distinct.

Among the perplexing similarities in both sloth families is the fact that they evolved from once-small and inconspicuous foragers into massive animals. The typical pattern of animal development – following Cope's so-called rule – is often from a smaller creature to a larger one, and there can be no doubt, although we have almost no trace of them, that there were tiny prototypical sloths who served as ancestors to contemporary species. What is striking, though, is that the ancestral sloths that we do know about and that are more than adequately represented in the fossil record were gigantic creatures which, at close to 4,800 kilograms (9,000 lb) and 6 metres (19 ft), were up to 350 times larger than their descendants. How this happened is not entirely clear, but it is yet another of the many perplexing turnabouts in the history of palaeontology.

For at least two centuries, well before early palaeontologists unearthed fossil giant ground sloths, the sloth was simply known as a diminutive, perhaps even contemptible, creature. Neither the two-toed nor the three-toed sloth weighed much more than 6 kilograms or, at most, 10 kilograms (14–22 lb), nor was there any sense that their predecessors were much larger or markedly different in constitution. That dramatic misconception was challenged as explorers and the anatomists of the late eighteenth and early nineteenth centuries, including the young Charles Darwin, stunned the scientific world by introducing massive and ungainly fossils of sloth-like creatures.

The discovery of giant ground sloths in the late eighteenth century created a flurry of interest with respect both to the sloth family and to fossils in general. The fascination with reptilian ancestors – that is, actual dinosaurs – would not take hold until the late 1820s, when the discovery of a fossil plesiosaur by Mary Anning (1799–1847) and the reconstruction of an *Iguanodon* by Gideon Mantell (1790–1852) captured the attention of the British public. And so, when an Argentinian fossil was discovered in 1788 and shipped off to Spain, Georges Cuvier, France's great comparative anatomist, took notice. This one fossil provided startling evidence that giant creatures roamed the planet in prehistoric times, and Cuvier, having determined that the fossil was an ancestor of the sloth, gave it the name *Megatherium americanum*. At roughly the same time, Thomas Jefferson, having been sent the fossil remains of a giant sloth by Colonel John Stuart, named that creature *Megalonyx* (great claw) after the oversized nails on its front feet. He described it in 1797 at a meeting of the American Philosophical Society in Philadelphia, and two years later, Jefferson published his work in *Transactions of the American Philosophical Society* with the title 'A Memoir on the Discovery of Certain Bones of a Quadruped of the Clawed Kind in the Western Parts of Virginia'.[3] The specimen was described in detail by the Philadelphia anatomist and physician Caspar Wistar (1761–1818), but it was eventually named after Jefferson – as *Megalonyx jeffersonii*

Composite representation of ice age mammals in what is now Patagonia, featuring *Megatherium* in the centre. Nearby animals include the elephant-like Stegomastodon, early armadillos, sabre-toothed cats, early horses and the llama-like Macrauchenia.

– by French zoologist Anselme Gaëtan Desmarest (1784–1838), although he applied the genus, *Megatherium*, rather than the more appropriate *Megalonyx*. Jefferson was optimistic that some of the large mammals he had encountered in fossil form still roamed the still-distant American West and hoped that the Lewis and Clark expedition might discover an enclave of these formerly extinct mammals. Meriwether Lewis was even given a crash course in anatomy by Caspar Wistar, but no 'living fossils' were found, notwithstanding the overall success of the expedition.

The giant claws of *Megalonyx*, which prompted Jefferson to mistakenly entertain the idea that his sloth was a lion-like predator, were no less big in *Megatherium*. And while most researchers still believe that *Megatherium* was predominantly herbivorous, its claws were impressive indeed. The size of these nails or claws recently prompted R. A. Fariña and R. E. Blanco, a palaeontologist and a physicist respectively, to produce a scholarly paper with the thrilling title '*Megatherium*, the Stabber'.[4] They demonstrate that not only are the claws capable of doing great harm, but the *Megatherium* could thrust its arms out rapidly, allowing it to either kill or seriously wound other animals. The question is what other animals would approach so large a creature to be potentially stabbed, and moreover, since self-defence does not necessarily entail being carnivorous, whether there is any evidence that the victorious *Megatherium* would have eaten any part of its attacker. These are serious concerns that the authors acknowledge are difficult to answer, although they do suggest that if *Megatherium* were a true stabber, it would be, in their words, 'the largest land mammal hunter to have existed'.[5] That said, the prevailing view is that most of the giant ground sloths were tree and plant browsers, with a broader palate than modern sloths. There is some evidence that

Side-by-side depiction of *Megatherium* (and young) with hair, and hairless.

the Mylodonts were also predominantly grazers who relied on grasslands for a good part of their diet.

Another area of speculation, also advanced by Fariña, is whether *Megatherium* and some of its South American relatives were as furry as they are depicted in virtually all representations. Echoing the title of his earlier paper, Fariña suggests in '*Megatherium*, the Hairless' that so large a creature – if hairy – would tend to lose a lot of water to maintain its temperature.[6] And water, in the Pampas, where *Megatheria* abounded and perhaps originated, then and now, was a scarce commodity. We know that certain northern species, such as Harlan's ground sloth (*Paramylodon harlani*), were fur-bearing creatures because we have skin samples from them revealing both fur and ossicles, or bone plates, which probably offered the sloth protection from predators. But the northern sloth migrants may have needed fur to survive the cooler climate of North America, still feeling the chill of the ice age. The animals that thrived in South America, however, may have been impaired by an extensive covering of fur, particularly

Ground sloths (*Nothrotherium*) may well have been prey for sabre-toothed cats, as depicted in this illustration by Zdeněk Burian, 1948.

because their large bulk was more apt to retain heat. Overheating can be dangerous for very large animals, especially if they have any prolonged periods of activity. Polar bears, for example, can move quickly (up to about 40 kilometres per hour or 25 mph), but can't sustain that pace without overheating. But of course, the bears do chase prey, whereas the sloths, even if they were 'stabbers', did not. The idea of hairless ground sloths is certainly one worthy of consideration, but it seems unlikely that such a theory will ever be fully substantiated.

Megatherium was only one of many 'monstrous' mammals that Cuvier described, but it was perhaps the most exciting, particularly as it didn't resemble any contemporary analogues in the modern era (as, say, both the mammoth and the mastodon resemble the elephant). The oversized nature of its bones fascinated artists, who depicted giant sloths in a wide variety of settings, frequently with an elegant 'gentleman' nearby for contrast, certainly for size and perhaps for sophistication. A sarcastic and often sceptical poem, 'The Quadrupeds' Pic-Nic', published anonymously in 1840, hinted at the possibility that so-called 'gentlemen' and giant sloths shared a historical, and perhaps ungainly, similarity. In this case, the narrator, the geological 'Dr Mole', refutes Cuvierian theories and pompously traces his own ancestry back to the giant sloths.

> There was one Mr. Cuvier, who talk'd of the sloth,
> But to listen to nonsense like this I am loth;
> From the strength of their limbs, and the make of their
>   paws,
> From the shape of their bodies, and length of their claws,
> I am firmly convinced they're related to me,
> And to this all philosophers ought to agree.[7]

Megatherium illustration, from Simeon Shaw, *Nature Displayed in the Heavens, and on the Earth* (1823), vol. II.

With the expansion of colonial interest in South America, more 'philosophers', in other words natural philosophers or scientists, visited the continent, and with increased geological excavations, giant sloths seemed to multiply. Among the many debts we owe to Charles Darwin, who, while still in his twenties, travelled to South America on HMS *Beagle*, was the discovery of several giant ground sloths. While in Patagonia, Darwin came across the fossil remains of a giant animal, which he promptly sent to England's foremost anatomist, Richard Owen (1804–1892), who designated it a giant sloth and named it *Mylodon darwinii*. The fascination with giant ground sloths did not abate, and fossilized sloth skeletons found their way into hundreds of natural history museums.[8] Owen eventually published perhaps the definitive text of the time, *Memoir on the Megatherium; or, Giant Ground-Sloth of America*, in 1861, which solidified the place of giant ground sloths in Victorian consciousness.[9] Even before Owen's text, however, the appearance of giant sloths only added to the mystery of sloths in general. Despite the similarities of anatomical structure, particularly in the skull and the hind limbs, the connection between these massive animals and their contemporary, rather diminutive 'descendants' was, to say the least, unclear. One can understand the puzzle that confronted naturalists with respect to the relationship between fossil and contemporary sloths, particularly in the decades before the theory of evolution was a reality, and the puzzles haven't been entirely resolved. In fact, it is only recently that we have come to understand that, first, similarities between *Choloepus* and *Bradypus* must (as I have noted) be

Northern tamandua anteater, which is arboreal.

attributed to convergence, and second (and quite logically) that the two animals are quite different from each other.

It is possible that some of the smaller fossil sloths, such as *Neocnus toupiti*, which lived in Haiti, could climb trees successfully and had arboreal habits similar to the modern *Tamandua*, or 'lesser anteater'. The same mode of existence may have been true of the delightfully named genus *Hapalops* (gentle face), a small (1-metre-long, or 3 ft) and slender ground sloth from the Miocene. The physical ability to climb trees was probably advantageous to the smaller sloths, who were undoubtedly pursued by predators. In any event, they seem to have survived later into the Miocene than their larger relatives. Still, like the giant ground sloths, they were quadrupedal, perhaps using the knuckles of their forelegs for loco-motion, except when they raised themselves up to forage on trees or possibly when walking alertly for short distances. And like so many of the larger sloths, they were pedolateral – in other words, they walked on the outside bones of their feet. By contrast, humans are plantigrade, which means that we walk on a foot structure that runs from toe to heel, while many other animals (including dogs,

Megatherium clearly showing the pedolateral positioning of the feet, illustration by Carl Buell.

cats, horses and even elephants) are digitigrade, meaning that they have evolved to walk on their toes.

While virtually all sloths exhibited some form of pedolateral development, there was often variation in the skeleton more generally. The smaller-sized *Diabolotherium*, of the Megalonychid family, named for having been discovered in the 'Casa de Diablo' in the Andes, seemed to have lighter and more flexible forelimbs, which appeared to be well suited for climbing. They may well have climbed trees at lower elevations, while also being capable of scaling rocky surfaces in the Andes to find caves or simply to avoid predators. The ascent of these sloths in mountainous regions presents another interesting facet of their physiology: that they must have been capable of enduring cooler temperatures.

Another forearm development can be found in the much larger *Lestodon*, a Megatheriid, which had a pronounced elbow, allowing for the insertion of powerful muscles to facilitate digging. Although these sloths ate leaves, they would also turn over soil for vegetation and possibly roots. Fossil teeth from *Lestodon* are high-crowned, which means that they were well suited for the additional wear and tear that comes with ingesting grit and dirt.

Ample food sources were obviously critical for all sloth species, particularly the larger animals, and the scarcity of sustenance was a serious problem in arid and frozen climates. No less important were water resources, which not only supported vegetation but could replenish fluids in these immense animals. A recent discovery at Tanque Loma in the southwestern region of Ecuador uncovered the remains of at least 22 giant ground sloths of the species *Eremotherium laurillardi*, large members of the Megatheriid family. What is striking about this find by Emily Lindsey of the University of California, Berkeley, and Eric López of Universidad Estatal Península de Santa Elena is that these sloths – which varied in age – seem to have died of thirst or some sort of disease at a communal water hole. This may suggest, according to Lindsey and López, that the sloths travelled in herds, an interesting behavioural insight, especially for various species of *Eremotherium* which migrated as far north as Florida, Georgia and South Carolina.

As little as we may know about the behavioural habits of the giant ground sloths, we do have evidence of something that they left in their wake: avocados. The seed of the avocado, large as it is now, was even larger in its ancient form. Only the great megafauna, such as giant ground sloths and the elephant-like gomphotheres, were capable of eating an entire avocado, and the fact that avocados were eaten whole was the key to their distribution and survival. By passing through the gastrointestinal system, the seeds were not only triggered to flourish once 'deposited', but they were often expelled, given the relatively slow digestion cycle of the sloths, a good distance from the original tree. We thus owe a debt to the giant ground sloths (and other megafauna) for contemporary avocado culture. It goes without saying that the great extinction of megafauna should have led to the extinction of the avocado as well, but the survival of the fruit – which has gained lasting popularity – suggests that some human cultures had already begun to cultivate avocados on their own.

If we trace sloth evolution as far back as we can, there appear to be at least four (or possibly five) groups under the Xenarthrans including the Mylodonts, the Megatheriids, the Nototheriids and the Megalonychids. The Mylodonts may be the oldest and perhaps the most primitive of the four groups, many of which may have been active diggers, but what their objective was in digging isn't clear. *Mylodon* remains have been found in Chilean caves, but even that aspect of their lives isn't well understood. It is unlikely that caves were a refuge from predators, given the impressive size of the *Mylodons*, but they may well have provided shelter for raising young, even though we have little evidence of

Ossicles or dermal armour in the skin of Harlan's ground sloth.

their social behaviour. Although most species have been found in South America, *Paramylodon*, better known as Harlan's ground sloth, migrated to North America and is well represented among the many fossil finds in California's remarkable La Brea Tar Pits, as well as in the Diamond Valley Lake site in Hemet, California. One of the more distinctive characteristics of Harlan's ground sloth are the ossicles, or bone plates, that were distributed in the skin of the animal. The actual function of the ossicles, despite being referred to as dermal armour, is uncertain, which is exacerbated by the fact that we do not have a full pelt of sloth skin. A study of the structure and distribution of ossicles in Mylodontid sloths, by Néstor Toledo of the Museo de la Plata and his colleagues, allows for the possibility that ossicles offered some protection from predators in juvenile sloths and 'young and pregnant females'. There is, as they suggest, a lot more work that needs to be done, and much of what we learn will be contingent on future finds of mummified sloth skin.

We tend to think of sloths and even giant sloths as residents of South America, and for good reason: not only did sloths originate in South America, they thrived on the continent, and now can only be found there. But ground sloths had a long and successful history in North America once the continents were joined. The migration of giant ground sloths to North America was only possible because of the formation of the Isthmus of Panama about 3 million years ago. The forming of that bridge resulted in what has become known as the Great American Biotic Interchange (GABI, sometimes called the Great American Faunal Interchange, or GAFI), which allowed biogeographic traffic to travel both north and south in a dramatic redistribution of animals in the Americas. The sloths that made their way north and west were certainly successful, having migrated as far as California, Alaska and Canada. The La Brea Tar Pits in Los Angeles have excavated the remains of three sloths, in three separate genera: Jefferson's ground sloth (*Megalonyx jeffersonii*), the Shasta ground sloth (*Nothrotheriops shastensis*) and Harlan's ground sloth (*Paramylodon harlani*). Further north, fossil remains of Jefferson's ground sloth have been found in Alaska, Saskatchewan, Alberta, British Columbia, Yukon and the Northwest Territories. There is, in fact, a collector's coin, released in 2010 by the Royal Canadian Mint, that depicts *Megalonyx* grazing on what is undoubtedly a Canadian tree.

The success of the northern sloth migrations was, alas, limited. Ultimately, glaciation, combined with increasingly limited resources and perhaps even predation (whether from early humans or not), led to the disappearance of the northern sloth population. Their demise was probably in advance of that of their South American relatives, who, though equally doomed, would leave three-toed and two-toed descendants. 'Doomed' may not be the best choice of words here. First, because doom suggests a tragedy in the course of evolution rather than a progression of events that, following Darwin, is the ongoing state of life on earth. The impact of the ice age was serious enough that the giant sloths would likely have gone the way of all American megafauna, even if hunters did not target them as prey. Also mitigating against doom is the lingering, if wistful, hope that somewhere in the western forests of South America, *Mylodon* might still exist. This illusory whim has fuelled more than a few narratives in science fiction, but it did have some minor traction in the world of science. In the late 1890s the British explorer Hesketh Prichard (1876–1922) was sent by Sir Arthur Pearson, who owned the now-notorious *Daily Express* newspaper, to explore South America, ostensibly to search for a living *Mylodon*. On his return, Prichard wrote, in *The Heart of Patagonia* (1902, the work that resulted from his visit) that he 'could discover no trace whatever either by hearsay or from the evidence of our own experience to warrant the supposition that [the *Mylodon*] continues to exist to the present day'.[10] Still, Prichard felt compelled to offer a tiny shred of hope by reminding his readers how vast the country was, and thus that his search could not be exhaustive. That shred of hope was taken up by Britain's celebrated and prolific zoologist E. Ray Lankester (1847–1924) in his *Extinct Animals*, in which he tentatively and perhaps optimistically suggests that 'Possibly, but by no means probably, the Mylodon is still living . . . [in] region[s], as yet unvisited by man.'[11]

ACEDIA.

Subijce humerum tuum & porta sapientiam,
ne acedieris vinculis eius. *Eccli. 6.*

7

4 The Sin of Sloth and the Sloths of Sin

The slothful hate the busie active men.
And are detested by the same again.
Horace (65 BCE–8 BCE)[1]

It is easy to confuse the history of the idea of sloth with the natural history of the animal that we know as the sloth. The reader can be forgiven some degree of uncertainty about which version of the word appeared first in history: the noun or the adjective. The question is easily solved given that the animal was a zoological latecomer in the history of Western exploration, and the term 'sloth' dates back, as far as we know, to around 1175 as the word *slauðe*. This, needless to say, was well before sloths, the animals, were introduced to the Western world. This was in 1535, when the Spanish colonialist and historian Gonzalo Fernández de Oviedo (1478–1557) identified them with a sense of irony as *periquitos ligeros*, or the 'swift parrots'. Oviedo continued, much less ironically, by characterizing the sloth as 'the stupidest animal that can be found in the world . . . so awkward and slow in movement that it would require a whole day to go fifty paces'.[2] If we consider Juan Bautista Muñoz's 1535 illustration of a sloth, copied from Oviedo, its physical appearance simply insists on awkwardness.

It was the English botanist and anatomist Nehemiah Grew (1641–1712) who first used the term 'sloth', or, more accurately, 'sloath', to describe the *Bradypus*. It is, he wrote,

> an Animal of so slow a motion, that he will be three or four days, at least, in climbing up and coming down a Tree. And

Hieronymus Wierix, *Sloth* (*Acedia*), from the series *The Seven Vices*, before 1612, engraving.

to go the length of fifty Paces on plain ground, requires a whole day. The Natives of *Brasile* call him *Haii*, from his voice of a like sound . . . Whatsoever he takes hold of, he doth it so strongly (*or, rather stifly*) as sometimes to sleep securely while he hangs at it.[3]

Thus, given the immediate perceptions of Oviedo, Grew and others, the animal and the idea of sloth became ineluctably linked. Perhaps we need to look further back at the history of slothfulness to understand the human 'flaw' that became emblematized by this once-obscure animal.

Sloth, as a term describing laziness, indifference and ultimately sin, dates back roughly to the Christian monk Evagrius Ponticus (345–399 CE), who compiled a list of eight evil thoughts that would result in everlasting punishment. Numbered among these sins, which were subsequently reduced to seven in order to mirror the seven heavenly virtues, was *acedia*, a Latin term meaning 'without care'. *Acedia* was meant to suggest the spiritual indifference of indolence, but it gained currency as a term comprehending all

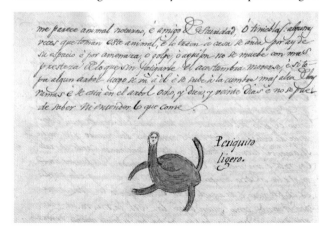

Juan Bautista Muñoz's copy of Gonzálo Fernández de Oviedo, *Historia general y natural de las Indias* (1535).

Detail from Pieter van der Heyden, after Pieter Bruegel the Elder, *Sloth* (*Desidia*), from the series *The Seven Deadly Sins*, 1558, engraving. In this image, snails and a donkey are the animal analogues of slothfulness.

forms of laziness. In early Middle English, roughly around the twelfth century, *acedia* gave way to *slāwðe*, the word for 'slow', and from it, the quality (or sin) that we call 'sloth'. Oddly enough, the earliest application of sloth in a zoological context dates back to the fifteenth century, well before sloths had ever been observed by Europeans. The word 'sloth' served as a collective noun that was applied to a group of bears, as in 'A Slouthe of Beerys'. The resulting modern phrase is essentially unchanged and remains: 'a sloth of bears', even though bears are not generally a very social group. The origin of the phrase is unclear, but surely has to do with the ambling and often unhurried gait that distinguishes bears in the wild.

In Latin, as well as in Spanish and Portuguese, the term *desidia* is used to describe slothful behaviour, as depicted, for example, in the Boschian image by Pieter van Heyden (1525–1569). Sloth appears in the allegorical figure of a drowsy woman, who is draped over an ass while the Devil adjusts her pillows, and a snail

87

moves its way out of the scene. There is, of course, no actual sloth in the image, but what is both fascinating and troubling is that when sloths took over as the iconic emblem of *desidia*, they couldn't escape being characterized as lazy, a term wholly inappropriate for an animal in the wild.

The sloth has existed in South and Central America for millennia and has been known under several names to Indigenous peoples, but most prominently as *ai*. The first published use of the word 'sloth' to describe the group was by the English compiler Samuel Purchas (1577?–1626), who wrote, 'The Spaniards call it . . . the light dog. The Portugals Sloth. The Indians, Hay.' [4]

Whatever the perception of the sloth was among Indigenous peoples, it rapidly became a symbol of idleness and indolence once it was identified by colonial explorers. Putting aside the symbolic connection with the so-called sin of sloth, it remains an infuriatingly inaccurate representation of the animal as a living creature. The term is exasperating because sloths are neither idle nor lazy, nor lethargic in the conventional sense of those words, but rather they are active *sloths*. If we mistakenly impose the connotation of the term on the animal itself, we are imposing Western cultural and religious implications on creatures that are, of course, indifferent to human value judgements.

Sloths, in keeping with all but the most domesticated animals, are oblivious to human expectations and interpretations; they simply live out their lives – at the rate they choose – in an environment that adheres to evolutionary principles rather than moral values. Nevertheless, early observers consistently mocked the animal for its supposed indolence and ineffectuality. The unwillingness to accept the sloth on its own terms can be disturbing, for example when Edward Bancroft (1744–1821), in his Guyana journal, tries to correct the 'insuperable aversion to motion' in sloths by 'beating' them, which results in some movement but only at the expense of 'the most melancholy pitiful noise and grimaces'.[5] Fortunately, the perspectives of other observers did have some sway. The renowned zoologist George Shaw (1751–1813), though hardly a sloth admirer, was willing to admit in his *Naturalist's Miscellany* (1789)

> that the Sloth, notwithstanding this appearance of wretchedness and deformity, is as well-fashioned for its proper modes and habits of life, and feels as much pleasure in its solitary and obscure retreats, as the rest of the animal world of greater locomotive powers, and superior external elegance.[6]

Shaw's view is surprisingly advanced for its time, though he still finds the sloth to be an object of 'pity and disgust'.[7]

The idea of slothfulness had its place in culture well before the arrival of the animal in Western zoological texts, and slothful characters abound in literature. Homer's lotus-eaters may have derived their slothfulness from their addiction to a narcotic plant, yet they set the tone for indolence for millennia. Alfred, Lord Tennyson, in his poem 'The Lotos-Eaters' (1832), draws on Homer but also suggests a certain malaise of his own – perhaps recalling the untimely death of his friend Arthur Hallam – when he writes, 'Surely, surely, slumber is more sweet than toil.'[8] St Thomas Aquinas argued against the status of sloth as a sin in his *Summa theologica* (1265–74) but considered it 'an oppressive sorrow, which . . . weighs upon man's mind, that he wants to do nothing . . . Hence sloth implies a certain weariness of work.'[9] Dante's *Purgatorio* offers a similar view of sloth in the form of two characters willing to succumb to despair, moral inertia and no aspirations to the betterment of themselves or others.[10] Geoffrey

Chaucer (1340–1400) was an astute reader of St Augustine, but for him – at least in 'The Parson's Tale' in *The Canterbury Tales* (*c.* 1400) – sloth ('Accidie [*sic*]') is very much a sin that requires tremendous resistance.

> For certes, ther bihoveth greet corage agains Accidie,
> lest that it ne swolwe the soule by the synne of sorwe,
> or destroye it by wanhope.

> For certainly, great valour is needed against Sloth,
> lest that it swallow the soul by the sin of sorrow,
> or destroy it by despair.[11]

The moral injunction against slothfulness is deeply rooted in Judaeo-Christian culture; in the Bible, for example, we are told that 'One who is slack in his work is brother to one who destroys' (Proverbs 18:9), which is one of several passages that has been consolidated into the familiar idiom 'Idle hands are the Devil's workshop,' or, as Isaac Watts warns his young readers in 'How Doth the Little Busy Bee' (1715), 'Satan finds some mischief still/ For idle hands to do.' Proverbs also suggests (6:6) that those who are lazy should go to the ants and learn from them, undoubtedly predating Aesop's (*c.* 620–564 BCE) advice in his fable 'The Grasshopper and the Ant'. The equally indolent Hare in Aesop's 'The Tortoise and the Hare' mocks his slow-moving rival, only to be defeated by his own idle habits. The idea that 'Slow and steady wins the race,' or the motto *festina lente* (used as a credo by Emperor Augustus (*c.* 27 BCE) and the Medici family (*c.* 1550s) and as a colophon for the great printer Aldus Manutius (1451–1515)) offers a different and more positive perspective on slowness (albeit not slothfulness). Caution in all things is not only advisable, but often necessary. Slow and gradual progress, unobservable to our senses, was (and remains) a centrepiece of natural history. As Darwin writes in *On the Origin of Species*, nature

> can produce no great or sudden modification; it can act only by very short and slow steps. Hence the canon of '*Natura non facit saltum*' [Nature does not make jumps], which every fresh addition to our knowledge tends to make more strictly correct.[12]

Notwithstanding the virtues of slow evolutionary and geological change, sloth as a personified quality, to say nothing of being a sin, has been the target of derision in literature. Lucifer, in Christopher Marlowe's *Doctor Faustus* (1620), makes a 'show' of the seven deadly sins, the sixth of which is sloth, who, beyond offering a brief self-introduction, is too lazy to 'speak another word'. The deadly sins are also characterized in various forms in *The Pilgrim's Progress* (1678) by John Bunyan; the prefatory poem, in fact, includes the author's recommendation that his book should serve as antidote to human indolence: 'This book will make a traveller of thee,' he writes, 'Yea, it will make the slothful active be.'[13]

The irony of the ever-productive Samuel Johnson naming his series of essays in *The Universal Chronicle* 'The Idler' should not be lost on contemporary readers. Johnson's social critique, perhaps reflecting a repressed side of himself, commends that slothful side of human nature that we all know. 'Idlers', Johnson writes, 'are always found to associate in peace; and he who is most famed for doing nothing, is glad to meet another as idle as himself.'[14]

One of the most interesting figures of laziness or slothfulness in literature is the 'wonderfully fat boy' Joe, in Charles Dickens's *Pickwick Papers* (1836–7), who is also marked by 'calmness and repose', or – in a word – sleepiness. Joe's 'condition' has been diagnosed in contemporary medical circles as obesity hypoventilation syndrome, a situation in which a person's obese condition prevents the adequate intake of oxygen. Known popularly as 'Pickwickian syndrome', the malady is serious and, of course, is unrelated to the behaviour of sloths except as an example of Dickens's disdain for Victorian indolence.

And in the canon of literary sloths, we cannot omit Herman Melville's *Bartleby the Scrivener* (1853), who devolves (as it were) from an energetic scribe to a pathologically apathetic character who

Joe, the 'wonderfully fat boy', from Charles Dickens, *The Pickwick Papers*, illustration from *Pictures from Dickens with Readings* (1895).

reacts to any request for work by stating simply, but emphatically, 'I would prefer not to.'[15] Bartleby's life, an extended exercise in the futility of nineteenth-century bureaucratic life, evokes the very modern interest in decorating the workplace with images of sloths.

Another intriguing, though lesser-known, sloth-like literary figure is the title character of Ivan Goncharov's novel *Oblomov* (1859), whose overwhelming apathy and slothfulness produced the term 'Oblomovitis' (обломовщина) to describe unremitting, in fact fatal, laziness. To return to Dickens, many readers may recall that in *A Tale of Two Cities* (1859), the apathetic Sydney Carton laments his 'sloth and sensuality', only to redeem himself at the conclusion of the novel. An idle man whose life has been devoted to resting when another is active, he is finally inspired to action because of his devotion to Lucie Manette. And few readers of literature are unfamiliar with his final words: 'It is a far, far better thing that I do, than I have ever done; it is a far, far better *rest* that I go to than I have ever known [emphasis added].'[16] Less heroic is Lewis Carroll's dormouse, one of the members of the Tea Party in *Alice's Adventures in Wonderland* (1865), who dozes off with alarming, albeit amusing, constancy. He clearly lives a slothful life that he justifies with Carrollian logic; when seemingly 'talking in his sleep', he tells Alice, 'You might as well say "I breathe when I sleep" is the same thing as "I sleep when I breathe!"'[17] Although his logic, by every measure of common sense, is flawed, it idiosyncratically holds true for the dormouse.

A later satirist, Evelyn Waugh (1903–1966), mixed his deft comic touch with an ever-present sense of disappointment in human behaviour. Along with Dante and Aquinas, Waugh recognized sloth, in its broadest sense, as a critical failing in humanity. 'The malice of Sloth lies not merely in the neglect of duty,' he wrote for a collection of essays on the seven deadly sins, 'but in the

refusal of joy. It is allied to despair.'[18] That feeling of despair char-
acterizes Guy Crouchback, the aptly named protagonist of
Waugh's trilogy about the Second World War, *Sword of Honour*
(1952–61).

One of the more charming reflections about sloths in fiction
appears at the outset of Yann Martel's *Life of Pi* (2001), when the
central character, Pi, reminisces about his zoological work with
sloths in Brazil. This prefatory reflection is in stark contrast to
the animal encounters that Pi experiences while shipwrecked in
a lifeboat with a tiger. 'The three-toed sloth lives a peaceful, vege-
tarian life,' Pi tells the reader, perhaps setting a peaceful backdrop
for the dramatic events to follow,

> in perfect harmony with its environment. A good-natured
> smile is forever on its lips . . . I have seen that smile with
> my own eyes. I am not one given to projecting human traits
> and emotions onto animals, but many a time during that
> month in Brazil, looking up at a sloth in repose, I felt I was
> in the presence of upside-down yogis deep in meditation
> or hermits deep in prayer, wise beings whose intense
> imaginative lives were beyond the reach of scientific
> probing.[19]

Pi's musing about sloth behaviour helps to ease him into a level
of comfort suggesting that a life of harmony with the natural
world remains possible, even in the aftermath of trauma.

Laconic, if not entirely slothful, characters abound in early
twentieth-century cartoons. Walt Disney's *Snow White and the
Seven Dwarfs* (1937), which was an interpretation of the Grimms'
fairy tale of 1812 as well as of a Broadway production in 1912, was
the original source of the allegorical names given to the Seven
Dwarfs. Even though the number 7 may bring to mind the deadly

sins, the dwarfs, aside from Sleepy, who comes closest to Sloth, do not represent the six other sins.

E. C. Segar created the character 'Wimpy' (also known as J. Wellington Wimpy) in 1931 to serve as straight man and foil to the hyper-masculinized Popeye the Sailor. Wimpy was not only sluggish and gluttonous but bore traces of the kind of shabby aristocracy that one might have expected from so distinguished a name in the midst of the Great Depression. In a sense, he was a parasitic character with neither the entrepreneurial self-confidence of Popeye, nor the brutal survivalism of Popeye's adversary, Bluto.

Although the name of the slightly later cartoon character 'Droopy' evokes slothfulness, he actually plays against that type. Droopy was an ostensibly weary and downtrodden cartoon basset hound created by Tex Avery (1908–1980), yet he was appealing because of his paradoxical success as a seemingly passive protagonist. Unlike Wimpy, Droopy always triumphed in his cartoon adventures, claiming to be, in as drowsy a voice as possible, 'happy'.

No doubt many more modern figures of slothfulness will come to mind for the reader, but a few over-the-top cinematic depictions deserve mention. George Lucas's Jabba the Hutt (*Return of the Jedi*, 1983) not only resembles an oversized slug, but is clearly immobile and physically passive even though he is an important (and menacing) intergalactic crime boss. Sloth makes a brief and grisly appearance in the movie *Seven* (1995), a crime thriller in which each of the victims represents one of the seven deadly sins. Much less intense is the Coen brothers' cult film *The Big Lebowski* (1998), which hinges, in part, on Jeff Bridges's portrayal of 'The Dude', an aimless, naive and directionless character who, after finding himself a victim of intrigue, is finally restored to a life of indifference. His own tag line, 'The Dude Abides', can be found on the same shelves as sloth T-shirts that proclaim a similarly apathetic 'Dudeness'.

Modern take on sloth by Tom Cheney from the *New Yorker*.

"This is sloth—greed is on the top floor."

Finally, the most absurd sloth-like character may be Gilligan, one of seven castaways from the 1960s sitcom *Gilligan's Island*, who, according to creator Sherwood Schwartz, was the avatar of 'sloth' in a cast that represented the remaining six sins. The clumsy and slothful Gilligan, for those who choose to remember the television show, was consistently the obstacle preventing escape from that island purgatory.

In the past decade or so, sloths have, for some reason, emerged as a cultural meme. Rather than uncouth, they are now typically represented as cute and endearing. And their so-called slothfulness has been co-opted and celebrated, particularly in North America, as an ideal for overworked individuals living in a society that prizes productivity above all else. T-shirts with sloth images abound, and the text on these shirts, though predictable (and zoologically inaccurate), is still amusing: 'Live Slow, Die Whenever', 'Born to be Mild', 'Sloth Running Team: We'll Get There When

We Get There'. The irony that lies behind these tag lines betrays something of the duality of sloths and slothfulness over the centuries. On the one hand, sloths were loathed and vilified for being emblematic of the kind of lethargy and indolence reviled both by the Church and subsequently by a capitalistic system that values efficiency and productivity. On the other, sloths are held up – rightly or wrongly – as emblems of a lifestyle that captures a fanciful vision of utopian leisure. In the final analysis, however, setting aside these highly imaginative views of the sloth, it is worth recalling that the very existence of the sloth underscores the impressive fact that these creatures can and do survive in nature by expending a minimal amount of energy and without requiring very much social engagement. In short, the sloth, whether in reality or in our imagination, has proven itself to be fascinating, loathed and loved, and ultimately compelling.

Ironic sloth T-shirt and meme.

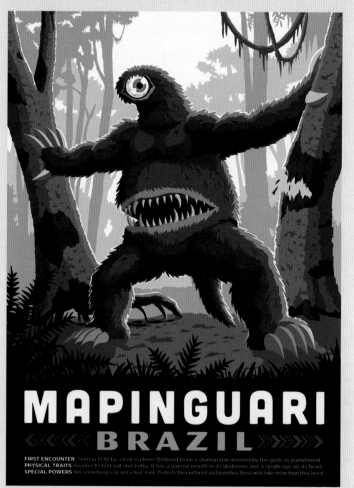

MAPINGUARI

BRAZIL

FIRST ENCOUNTER Seen in 1930 by a lost explorer. Believed to be a shaman transformed by the gods as punishment.
PHYSICAL TRAITS Around 10 feet tall and bulky. It has a gaping mouth in its abdomen and a single eye on its head.
SPECIAL POWERS Has a terrifying roar and a foul smell. Protects the rainforest and punishes those who take more than they need.

©2022 Anderson Design Group, Inc. All rights reserved. www.ADGstore.com

5 Sloths in Culture

The Nimrods of the nursery will ride nothing
but a Megatherium.
Punch magazine, 1848

Wisely, and slow. They stumble that run fast.
William Shakespeare, *Romeo and Juliet*, Act II

It might be argued that the sloth has become the most emblematic animal from South and Central America. Notwithstanding the presence of exciting and charismatic fauna such as jaguars, condors, anacondas, coypus, capybaras and even the domesticated llama and alpaca, no animal evokes the region as much as the sloth. This, no doubt, has a great deal to do with tourism and marketing strategies that foreground the normally reclusive animal. In the wild, sloths can be elusive, and given that *Choloepus* is nocturnal by nature, it isn't as if these creatures by habit (they are not social in any way) or by appearance (they are very well camouflaged) call a great deal of attention to themselves.

The fact that sloths are remarkable examples of one of the more radical shifts in size in evolutionary history is, by any standard, impressive. Yet even though this transition has captured the fascination of every trained zoologist and palaeontologist, it is doubtful that the phylogeny of sloths (however unique) is a major factor in general sloth appeal. What electrifies palaeontologists and zoologists doesn't always excite the public, and the eccentric prehistory of sloths surely doesn't explain current trends. What *does* fascinate tourists and contemporary naturalists alike is the slowness of these animals in almost every aspect of their lives. For most people, the sloth is an iconic embodiment

Poster depicting
the Mapinguari
by Nashville's
Anderson Design
Group. With four
toes and five
fingers, the image
distances itself
from the giant
ground sloth.

of justifiable passivity, in contrast to the apparent freneticism of other animals and, needless to say, human society.

Many of us, readers of this book included, are residents of intense industrial cultures, in which there is a high premium on being productive and efficient. In order to shrug off that intense workload, we have made a fetish of the sloth, which, for us, represents an enviable lethargy and indolence. The serene, almost comforting, appearance of the genial sloth is an appealing image that reflects an indifference to any need to be energetic or to hustle for one's living, and so it is not surprising to find sloth paraphernalia with catchphrases like 'time to get slothed', 'you can't rush perfection', 'life in the slow lane' and 'no hurries, no worries' adorning tens of thousands of leisure products, from figurines to T-shirts. These quips, of course, have little to do with sloths themselves and everything to do with a culture that wants to read a perception of 'slothiness' into their own lives, perhaps with more than a tinge of irony.

The cute appeal of sloths, it should be noted, is a very recent phenomenon. The age-old perception of sloths as inept creatures is borne out even in relatively recent decades, for example *All about Strange Beasts of the Present* (1957), a book for young adults in

Souvenir sloths appear as rubber sloth ducks and in snow globes.

which the once-prominent science writer Robert S. Lemmon notes: 'Sloths have the world's dumbest faces. Their small, weak, dull gray-brown eyes show no expression whatever. The brain behind them is as stupid as they are.'[1] The remainder of Lemmon's description is equally heartless and, more to the point, inaccurate. One of the darlings of his book – notwithstanding its often-aggressive temperament – is the 'appealing little koala', a 'natural-born teddy-bear'; koalas were somehow considered fashionable animals – for reasons that aren't clear – in the 1950s and '60s.

Central and South America were plagued by violent political tensions until late in the twentieth century, which limited their appeal to tourists, and thus sloths were less prominent than they might have been until the last few decades. Nearby countries where sloths can be found, including Venezuela, Bolivia and Brazil, have, in the last decade or so, faced ecological challenges due to conservative, repressive and autocratic regimes. In many of these countries, the appeal of eco-tourism, to say nothing of sustainability, has given way to the relentless push for economic expansion.

Costa Rica, which itself experienced a divisive and bloody civil war in 1948, has managed to reinvent itself, over time, as a stable democracy. Under the leadership of José Figueres Ferrer (1906–1990), Costa Rica adopted a new constitution and also abolished its army in 1949. The country is thus an exception to the rule of regional instability, and it remains one of the most important tourist destinations in Latin America. Costa Rican holidays may not always be about sloths, but surely no visitor can escape the ubiquitous images of sloths; most tourists in Costa Rica will, in fact, pay for sloth merchandise with a banknote that prominently displays a three-toed sloth.

In addition to the proliferation of sloth images on an astonishingly wide array of tchotchkes, beach towels, sweatshirts, socks and so on, actual living sloths in the wild, in captivity and even in

Costa Rican 10,000-colon banknote featuring a three-toed sloth.

private hands continue to captivate the public. Sloths contribute substantially to eco-tourism in Central and South America, particularly in Costa Rica, where there are several sloth sanctuaries, although they can also be found in Peru, Honduras and Colombia. The sanctuaries, in general, perform a valuable function with respect to sloths and other indigenous animals, but they also have a zoo-like, if not actually touristy, function for international visitors. The education (and entertainment) of tourists extends to national parks, where trained tour guides point out sloths, howler and capuchin monkeys, coatis, quetzals and toucans. For all of the excitement of seeing a sloth in the wild, the thrill doesn't usually last that long, and once a sloth has been spotted, it generally does not take much time until the tourists are prepared to move on to observe something else. Sloths, more specifically *Choloepus*, are becoming a little more common in zoos and aquaria throughout the United States, where there may be as many as two hundred animals distributed across the country.

The more disturbing question is how many sloths are actually in private hands in the world. The three-toed sloth does not survive well in captivity, partly because of its very limited diet and its susceptibility to illnesses ranging from pneumonia to fungal infections. In addition to the illnesses they are subject to, sloths in captivity require a great deal of care and have important feeding restraints. Most countries do not have very restrictive laws concerning the sloth in particular, although laws dealing with wildlife trafficking in general may be on the books. With habitat destruction so prevalent, poachers often acquire sloths from individuals selling 'found' infants, which, in one way or another, have been separated from their mothers. The cost for these 'orphans' can be as low as £5–£30. Within the legal exotic pet trade, a sloth may be purchased – at this moment – for roughly £9,000, to say nothing of maintenance and upkeep; undoubtedly, many privately owned sloths, if they do survive, are eventually given to zoos. Unfortunately, many animals don't even survive to market (as it were) because their sharp nails are often mutilated and, in *Choloepus*, their teeth are filed down, which not only impairs these sensitive animals from behaving naturally but imposes often-fatal stresses on them. The shameless practice of wildlife trafficking, requiring poachers as a primary source, is a global problem for thousands of species and is built on a foundation of affluent and reckless so-called collectors whose money will sustain these criminal activities until, due to extinction, there are no more living specimens of a particular species to collect.

With respect to observing sloths in their natural environment, there is still hope that the growing interest in and the affection for sloths will lead to greater habit preservation and protection. At this point, however, the renewed attention of Western culture has focused on entertainment and tourism rather than critical issues such as habitat preservation. The dilemma, which is familiar to every preservationist, requires not merely human admiration, but genuine engagement with the welfare of the very animals we admire.

It is interesting that sloths, in general, either because of their visual inaccessibility or their slowness, have not generated a strong local mythology. Sloths do not seem to have a robust presence in Central and South American folklore, perhaps because they never had considerable value as a food source for native populations and were certainly never a threat. One exception, in Brazil, is the legendary bipedal creature Mapinguari (a Portuguese name), which has many sloth-like features.[2] The animal, called *Owojo* (people eater) or *Kida harara* (laughing beast) in the native Tupi language, is not only large but dominated by a single eye and a vertical mouth centred where its stomach should be. Often large and red-haired, it is said to emit a foul smell that can be overwhelming, as well as being able to produce a disturbingly loud cry.

Whether this nightmarish cryptid is some form of residual memory of the giant ground sloths, which were no doubt frightening, or an amalgam of forest creatures is not clear. While some cultural memory of giant sloths cannot be discounted entirely, it

Rock painting showing a giant ground sloth and its young in the presence of six people, Colombia.

seems doubtful that it would be so resilient as to persist in the form of a mapinguari. Nevertheless, at least one image of a giant ground sloth, with its young, has been preserved in remarkable rock paintings – dating back about 20,000 years – by Palaeolithic residents of what is now the Chiribiquete National Park of Colombia.[3] And, from the perspective of myth and legend, there are persistent notions that somewhere in the deep forests or jungles of the Americas, some form of giant ground sloth still exists. Recall that Thomas Jefferson, fascinated with his own description of *Megalonyx*, advised Meriwether Lewis and William Clark in 1803 to be on the lookout for living specimens while on their westward expedition.

Isabel Allende's *City of the Beasts* (*La ciudad de las bestias*, 2002) invokes the myth of Mapinguari-like creatures, even down to their foul smell. These are creatures revered by the 'People of the Mist' but are, in fact, a hidden colony of intelligent and congenial ground sloths. The two child protagonists, Nadia (daughter of a South American guide) and Alex (grandson of an American reporter for *International Geographic Magazine*), are guided by a shaman to meet the creatures, and they are the only ones who are able to report what they have seen. Allende, in drawing on alienated children who are 'special' enough to experience the greatest mysteries of the Amazon, has adopted familiar motifs for her narrative, but the setting is distinctive, and the 'beasts' are, like the giant ground sloths themselves, appropriately enigmatic.

A similar but less fanciful figure than the Mapinguari is the creature known as the gran bestia, which was also feared, or at least respected. Charles Darwin noted the myth while fossil-hunting in South America. 'In connection with the Megatherium', he wrote in his 1832 geological diary, 'I may mention a curious fact. – It is a common report in all these parts of S America that there exists in Paraguay, an animal larger than a bullock, & which goes by the name of "gran bestia".'[4] The captain of the *Beagle*, Robert FitzRoy, was also told of a mysterious creature called a gran bestia,

> not many months old, but which then stood about four feet [1.2 m] high. It was very fierce, and secured by a chain. Its shape resembled that of a hog, but it had talons on its feet instead of hoofs; the snout was like a hog's, but much longer. When half-grown, he was told that it would be capable of seizing and carrying away a horse or a bullock. I concluded that he must have seen a tapir or anta [tapir in Portuguese]; yet as he persisted in asserting that the animal he saw was a beast of prey, and that it was extremely rare.[5]

Darwin was certain the gran bestia must have been based on the lingering legend of an animal conceived from the fossil remains of giant ground sloths. If that was the case, however, it would suggest a cultural memory going back more than 6,000 years at the very least, which, though not impossible, once again seems unlikely.

The mythology of the Amazonian peoples is not all that well recorded, although some legends about sloths, beyond the Mapinguari, persist. Claude Lévi-Strauss documents the myth that if a sloth were to defecate from above (rather than descend to the ground to perform the task), 'the excrement would change into a comet', which might shatter the earth.[6] The Shuar (sometimes called the Jivaro) were well known for the practice of *tsantsa*, the 'shrinking' of the heads of their enemies, but sloths were generally protected from this practice, because to do so would incur the curse of the sloth.

Sloths were not often hunted as bushmeat, no doubt because they are so small and thus carry very little muscle for consumption.

In W. H. Hudson's powerful novel *Green Mansions* (1904), a sloth – initially a source of tender companionship for the male protagonist, Abel – is abruptly killed for sustenance as Abel desperately attempts to leave the forest.[7] Necessity, among tribes like the Hi'aiti'ihi (also known as the Pirahã) of Brazil, may result in the eating of sloths, but local taboos keep this to a minimum. One rationale behind the taboo may well be that sloths, in leaving excrement on the ground, provide fertile sites for trees, which will eventually sustain more monkeys – themselves considered legitimate prey. The mythical female sloth creature Uyush is, in one tradition, considered the only source of manioc (cassava), and so human women were obligated to appeal to her deferentially to have access to the fruit. Furthermore, although not thought of as a desirable source of food, sloths do have a role in folk medicine: for example, the rendered fat of sloths was, in some cases, considered medicinally useful for spider bites and scorpion stings.

SLOTHS IN LITERATURE

The sloth has long been a potent icon of 'do-nothingness'. In one satirical cartoon by the great American editorial cartoonist Thomas Nast (1840–1902), corrupt and self-serving politicians were skewered in a depiction of elected officials in the form of a sloth, and the electorate in general as industrious beavers. The cartoon, 'Citizen Beaver and "Statesman" Sloth', which appeared in *Harper's Weekly* in May 1881, makes Nast's point eminently clear in the caption: 'While the Citizen Beaver labors for prosperity and good government, the "Statesman" Sloth lies on his back and obstructs and destroys.'

An equally ominous sloth makes a brief but ghastly appearance in H. G. Wells's *Island of Dr Moreau* (1896), as one of the

HARPER'S WEEKLY.

JOURNAL OF CIVILIZATION

Vol. XXV.—No. 1273.]   NEW YORK, SATURDAY, MAY 21, 1881.   { SINGLE COPIES TEN CENTS. $4.00 PER YEAR IN ADVANCE.

Entered according to Act of Congress, in the Year 1881, by Harper & Brothers, in the Office of the Librarian of Congress, at Washington.

CITIZEN BEAVER AND "STATESMAN" SLOTH.

While the Citizen Beaver labors for prosperity and good government, the "Statesman" Sloth lies on his back and obstructs and destroys.

assemblages of animals created by Moreau. Prendick, the protagonist of the novel, has quite a start when he is approached by 'a dim pinkish thing, looking more like a flayed child than anything else in the world. The creature had exactly the mild but repulsive features of a sloth, the same low forehead and slow gestures.' The sloth is one of many 'Beast Men' that Prendick ultimately encounters on the island. It is made clear that the 'sloth creature' has an exceptionally ugly face, yet he is also the individual who gently alerts Prendick to the truly menacing creatures around him.

In keeping with the growing interest in adventure tales in the nineteenth and early twentieth centuries, the Scottish-born physician and adventure novelist William Gordon Stables (1840–1910) published *In Quest of the Giant Sloth* (1902), a novel in the spirit of Jules Verne, H. Rider Haggard and later Arthur Conan Doyle.[8] The search, in a South American region the characters call Marvelland, leads to the discovery of giant sloths, and to Stables's credit, he resists turning them into monsters. Yes, they are immense, but Stables draws on subtlety rather than theatricality to describe them. The sloths, Stables writes, are

> partly bear, partly baboon; the face black and almost destitute of hair; a curious ruff and tippet around the neck of a dark, grizzled-gray colour, the hair on the body being exceedingly long and brown. But the eyes were fierce and terrible, and when it stood on end, with forefeet in the air, and uttered a prolonged and mournful moan, a more fearsome animal surely never existed out of a maniac's dream.[9]

The encounter between the sloth and the main child character, Charlie, is also prescient and touching, suggesting Stables's own concern for the extinction of what must have been a remarkable animal. The sound of the creature's 'voice', the narrator explains,

Thomas Nast's cartoon, 'Citizen Beaver and "Statesman" Sloth', on the cover of *Harper's Weekly*, 21 May 1881.

'was like the despairing piteous cry – if ever such a cry was heard – of some lost souls departing for eternal woe'.

A much less sympathetic and certainly less accurate view of giant ground sloths can be found in the 1948 adventure film *Unknown Island*, which is itself almost unknown if not entirely forgotten. The island in question is home to a colony of dinosaurs, mostly *Tyrannosaurus rex*, but for some reason the animal population also includes a bizarre-looking giant ground sloth. Outfitted in reddish fur and with a toothy (almost gorilla-like) muzzle, the costumed actor – almost certainly Ray 'Crash' Corrigan (1902–1976), who was famous for his many gorilla roles – had very little instruction, if any, to prepare him to be a *Megatherium*.

Front cover of Gordon Stables, *In Quest of the Giant Sloth* (1902).

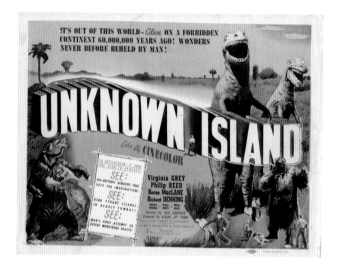

Poster for the adventure film *Unknown Island* (1948, dir. Jack Bernhard) depicting a conventional giant sloth at the lower left and the costumed version used in the film at the lower right.

In the final moments of the movie, a terrified couple watch as the sloth wrestles with and viciously bites a *T. rex*, and then manages to push it over a cliff to its death. The sloth then conveniently disappears to allow for a romantic denouement. It is, without question, a ludicrous and almost-unwatchable film, but it deserves note for making an effort to promote the *Megatherium* into a legitimate monster of the silver screen. Though the film was an abject failure, the very fact that the screenwriters decided to include a sloth in a dinosaur film, quite inappropriately and perhaps impulsively, is a curious juxtaposition of ostensibly real, but certainly not contemporary, prehistoric 'monsters'.

Less menacing *Megatherium* figures pop up, every now and then, in works for children. In *Five Children and It* (1902) by Edith Nesbit, for example, the young heroes of the novel learn from an ancient 'sand fairy' called a Psammead of a time when you could enjoy 'megatheriums for breakfast'. There were never any leftovers, however, because their bones turned to stone as soon as the meal was done. Thus, Nesbit uses the Psammead to point to contemporary fossils as an indication of past ages.

To return briefly to the historical accounts of sloths, it is always worth recalling that in Western eyes, sloths have been considered enigmatic ever since the earliest accounts of their existence. Like so many creatures when first seen by explorers, soldiers and missionaries, they were misunderstood and misinterpreted in their natural context. Oviedo, as we have seen, tried to make sense of this odd addition to his repertoire of animal knowledge, but reconciling so bizarre a creature with his experience was undoubtedly challenging. Oviedo, for example, never seems to have witnessed sloths in the act of eating, and so he believed that they existed on a diet of air. 'No one can find out what this animal eats,' he writes:

> I had one in my home, and from my observations I have come to believe that this animal lives on air. The sloth has never been seen to eat anything, but it turns its head and mouth into the wind more than any other direction, from which one can see that it is very fond of air.[10]

It isn't clear what possible foods Oviedo kept in his home, but he surely didn't have *Cecropia* leaves or hibiscus petals, which almost certainly would have stimulated the appetites of his sloths. But the belief that sloths might simply consume air is a notion that persists in several early accounts of the animal. Clearly, the feeling was that since the sloth was so apparently lethargic an animal, it needed very little sustenance (if any) to fuel its minimal activities. The early cleric and explorer André Thevet, who visited Brazil in 1555 and subsequently published *Les singularitez de la France antarctique* in 1557, also seemed convinced that sloths could live on a diet of air. His work provides perhaps the first published Western illustration

Drawing from
André Thevet, *Les
singularitez de la
France antarctique,
autrement nommée
Amérique* (1558).

of a sloth, although its human-like face (and ears), thick and hairy body, and quadrupedal stance bear little resemblance to the actual animal itself. The illustration is consistent with later depictions, however, in that the sloth seems amiable, with very human-like facial features. That human 'look' is certainly behind a very long history of anthropomorphized sloth images that continue to proliferate on calendars, T-shirts and souvenirs.

An equally compelling image of a sloth can be found in an illustrated work titled *Histoire Naturelle des Indes*, in what is known as the Drake Manuscript. The manuscript, written in French and part of the collections of the Morgan Library in New York, is dated around 1586 and may well have been the zoological and ethnographic work of French Huguenots travelling with Francis Drake. The animal depicted is a two-toed sloth (notably two-toed on the hind limbs as well), suspended somewhat awkwardly – though accurately – in a tree. 'The nature of this animal', the text observes,

Illustration of a
two-toed sloth in
*Histoire Naturelle
des Indes,* known
as the Drake
Manuscript,
c. 1586.

PERIQITE·LEGERE·

is to climb with its belly uppermost and, climbing this way,
it moves faster than a man could on foot. The skin of this
animal is very excellent for people suffering from the falling sickness [epilepsy]. The head of the afflicted is covered
with it and then one realizes how effective the skin is.[11]

Perhaps the sloth's inverted yet stable posture offered some hope
for individuals dealing with epilepsy, but we don't see evidence
of this 'cure' repeated elsewhere in sloth lore. While this appears
to be one of the only records of sloths as folk remedy, the sloth
recently served – perhaps coincidentally – as a mascot for the
Epilepsy Foundation.

Later representations of the sloth were slightly more accurate, including the bear-like creature in Johannes Stradanus's drawing 'Allegory of America', which includes a portrait of Amerigo Vespucci and was prepared for an Italian patron. While Vespucci dominates the image, a sexualized depiction of conquest and domination in the form of an Indigenous woman who is recumbent, submissive and seemingly powerless clearly represents newly colonized America. In the illustration, the artist also offers an array of native animals including an anteater just below the tree on the right-hand side, and, just above it (ascending the tree), a seemingly innocuous and affable sloth.

Despite the friendly illustrations and even beneficial descriptions of the animal, the denigration of the sloth is a constant theme of early natural history writing, and indeed has dogged the sloth until the last decade or so. The Comte du Buffon (1707–1788), whose estimation of New World animals was low indeed, singled the sloth out for his harshest criticism in his comprehensive *Natural*

Theodor Galle after Johannes Stradanus, *Allegory of America*, c. 1600, engraving. A sloth is climbing up the tree on the right side of the illustration.

History of 1760: 'One more defect', he pronounced, 'and they could not have existed.'[12] Buffon, either unwilling or unable to understand the creature in its context, assumed that the very features that allow the sloth to survive are liabilities to its existence. The sloth's 'uncouth conformation', he writes, results in 'slowness, stupidity, and even habitual pain'.[13] The great Buffon, as much as he is to be admired, had strong opinions that, alas, were not always sympathetic to ethology, much less to adaptive fitness. Sloths, of course, are neither stupid nor in habitual pain, but they are considerably slower and less social than most of the mammals with which we are familiar. Nevertheless, Buffon's judgement endured for close to a century, and sloths remained poorly understood, if understood at all, by the public.

The early modern reception of sloths may seem, in retrospect, wilfully contrarian, but it is worth recalling that the sloth was unlike any other creature that explorers had ever observed or recorded. Its slowness and upside-down existence could not yet be processed through the lens of what is now called adaptive fitness, and thus those traits were considered corruptions of the standard mammalian model. The sloth was certainly a perplexing (if not disconcerting) example of providential design. The earliest explorers exacerbated the situation by deliberately placing sloths on the ground ostensibly to position them like 'normal' quadrupeds. The resulting failures, which made the sloths seem only more pathetic (a term still used even by experts), resulted in a persistent misconception in zoological literature.

A very unnerving image of a *Bradypus* appears in the many editions of *The Natural History of Quadrupeds* (1790), by the remarkable engraver Thomas Bewick. Drawn in a horizontal climbing position, the animal not only has a severe look of consternation but appears to have very menacing canine teeth. Described in the text as 'helpless and wretched', we also learn that it wards off

MAMMALIA.

Order. Edentata.

Genera Bradypus.Dasypus.Myrmecophaga.Manis

Bradypodidæ.
Families Dasypodidæ
Myrmecophagidæ

D. Peba . Peba.

B. Tridactylus. Three-toed Sloth.

Myr. Jubata . Great Ant eater.

M. Tetradactyla . Long-tailed Manis.

predators by 'a most pitiable and lamentable cry', which instils 'horror' in animals that threaten it. The two-toed sloth is mentioned briefly, with the suggestion that it is also found in Ceylon.

Bewick's image is fitting for its time. As we have already seen, the very suggestion that any animal might live upside down was certainly confounding to early naturalists, and time and again, the sloth is represented as an upright quadruped. One of the standard texts in natural history was *Nature Displayed in the Heavens, and on the Earth, According to the Latest Observations and Discoveries* (1823) by Simeon Shaw. Shaw (who, perhaps, had never seen a live sloth) described them as follows:

The appearance of this animal is so helpless as to excite compassion. It has three claws upon each foot, and a short tail. Its fur is long and coarse, resembling dried grass; the

The three-toed sloth neatly curled up with its other Edentate (Xenarthran) relatives, engraving by F. W. Lowry from *Illustrations of Zoology* (1851).

Thomas Bewick's
woodblock and
impression of an
apparently angry
three-toed sloth,
c. 1790–1804.

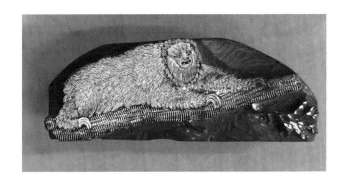

mouth extremely wide; the eyes dull and heavy; and the
legs and feet set on so awkwardly, that a few paces often
require the effort of a week. The legs indeed proceed from
the body in such an oblique direction, that the sole of the
foot seldom touches the ground.[14]

Shaw clearly understood that sloths like to climb and to suspend
themselves, but he nevertheless provides an image of a three-toed
sloth walking on all fours.

Representations of quadrupedal sloths persisted well into the
early nineteenth century in images that, at least for modern viewers, defy credulity. In an illustration supervised by the engraver
Robert Bénard, which appears in *Encyclopédie méthodique par*

118

ordre des matières, published between 1782 and 1832 by the French publisher Charles Joseph Panckoucke, the two-toed sloth is on all fours, but the three-toed sloth is depicted sitting on his haunches in a casual upright, almost simian, posture. These illustrations suggest an ongoing confusion about both the appearance and the posture of sloths. Still, despite the prevalence of overly imaginative, if not ludicrous, illustrations, Charles Bell, an eminent professor of anatomy and surgery at the Royal College of Surgeons in London, wrote soberly and pragmatically about the sloth in his *Treatise on the Hand* (1833):

> The sloth may move tardily; his long arms and preposterous claws may then be an incumbrance; but in his natural place, among the branches of trees, they are of advantage in obtaining his food, and in giving him shelter and safety from his enemies.[15]

Simeon Shaw's ungainly sloth walking on all fours in his *Nature Displayed . . .* (1823), vol. IV.

L'Unau, le $\hat{A}$ le

Papio . 2

Cercopithecus.
Meer Katz.

Charles Reuben Ryley illustrated this stuffed specimen of a two-toed sloth, c. 1792, which was in the eclectic collection of Ashton Lever (1729–1788).

Two apparently well-shorn sloths walking on all fours. Colour drawing by Victor Adam, 1853.

A representation of a three-toed sloth (called Papio, or 'baboon') in John Jonston, *Historia naturalis de quadrupetibus* (c. 1653).

Bell's thoughtful reflections about the sloth bolstered the legitimacy of sloths as a rational part of the animal creation; of course, Bell not only sustained the idea of sloths as 'normal' using his skills in anatomy, but validated them by finding (and restoring) evidence of God's wisdom and sense of perfection in every organism.

Even as modern sloths continued to intrigue both professional and amateur naturalists, the discovery of fossil giant ground sloths in the late eighteenth century created a new wave of interest in and excitement about the sloth family. An Argentinian fossil, unearthed by Manuel Torres in 1788 on the banks of the Luján River near Buenos Aires, created quite a sensation when it arrived in Madrid and was assembled and mounted for display in 1795 by Juan Bautista Bru (1740–1799), of the Royal Cabinet of Natural History. Bru's sloth is arranged in what might be considered a slightly awkward straight-backed quadrupedal pose, particularly since we have become so familiar with sloth displays that show *Megatherium*-like creatures leaning, in a bipedal posture, against a tree. But his work was remarkable, and his story deserves a little digression because it offers a fascinating example of politics and intrigue in early palaeontology.

Juan Bautista Bru's fascinating, albeit awkward, reconstruction of a *Megatherium* skeleton, 1793.

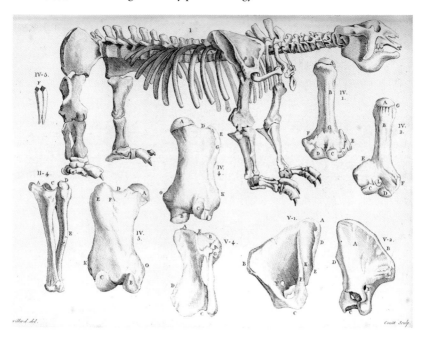

The historian of science José M. López Piñero has taken great pains to trace the career of Bru, who not only put the sloth together (no simple task for an unprecedented species) but compiled a text replete with his own illustrations that was titled *Description of the Skeleton in Detail, Following Observations while Mounting and Erecting It in the Royal Cabinet*. Unfortunately for the lesser-known Bru, copies of his as-yet-unpublished manuscript were obtained by a member of the Institut de France and passed on to Georges Cuvier, who named the creature *Megatherium americanum* in an essay published in the *Magasin encyclopédique* and later expanded into what became a classic text, *Sur le Megatherium* (1804). While it may be true that almost no one could match Cuvier's expertise and insight as a comparative anatomist, his connection to the *Megatherium* was only possible because of the now-long-forgotten Bru. To Bru's credit though, his assembled *Megatherium*, now slightly remounted and altered to conform to current knowledge, *still* stands – with all four feet on the ground – in Madrid's Museo Nacional de Ciencias Naturales.

A highly romanticized image of sloths appeared in *Narrative of a Five Years' Expedition* (1796) by John Gabriel Stedman. Stedman, although part of the Dutch colonial forces, recognized the abuses being perpetrated on the subjugated peoples of Surinam. That recognition manifests itself in his depiction of the 'sheep sloth' (ai, three-toed) and the 'dog sloth' (unau, two-toed), who take refuge on a blasted tree in a ravaged landscape. And so, the images of sloth could be emblems of social and political concerns even as early as the eighteenth century.

With the expansion of interest in South America, giant sloths seemed to multiply. During his trip to Patagonia, as we have already noted, Darwin came across the fossil remains of a giant animal, which he promptly sent to England's great anatomist Richard Owen, who pronounced it a giant sloth and named it

A highly romanticized engraving of sloths, characterizing the 'sheep sloth' (three-toed) and the 'dog sloth' (two-toed), from J. G. Stedman's controversial *Narrative of a Five Years' Expedition against the Revolted Negroes of Surinam* (1796).

Mylodon darwinii. Darwin went on to gain some renown for his discovery of *Megatherium* fossils. A letter from his good friend and correspondent the entomologist Frederick William Hope must have elated the 24-year-old Darwin, who was still voyaging on HMS *Beagle*. 'From sending home the much-desired bones of Megatherium', Hope wrote to the still-unknown Darwin, 'your name is likely to be immortalized at the Cambridge Meeting of naturalists; your name was in every mouth & Buckland applauded you as you deserved.'[16] The enthusiasm of Hope's letter can be forgiven if only because of his affection for Darwin and the fact that his surname was already widely known because of Erasmus Darwin (1731–1802), Charles's famous grandfather, a physician, celebrated poet and early evolutionary theorist. Many subsequent fossils were unearthed in South America, adding to Darwin's specimens, but fortunately, Darwin's own discovery of giant sloths undoubtedly contributed to his formulation of evolutionary theory. And so, in some small way, the *Megatherium* and its allies played at least a small role in the Darwinian legacy.

The Buckland mentioned by Hope was the Rev. William Buckland DD (1784–1856), who was well established as a theologian (eventually Dean of Westminster) and also a geologist and palae-ontologist. His 1824 account of the *Megalosaurus*, which makes an appearance in Charles Dickens's *Bleak House* (1853), was one of the first descriptions of what would eventually come to be known as dinosaurs. More important, here at least, is Buckland's interest in the *Megatherium*, which he discussed at length in *Geology and Mineralogy considered with Reference to Natural Theology*, his 1836 contribution to the Bridgewater Treatises. For Buckland, both leth-argic modern sloths and the ungainly *Megatherium* presented a problem in that they are typically described, by Buffon and others, as less than perfect organisms. This, according to Buckland, is an untenable argument if we are to believe in God's wisdom as a creator. 'It has been the constant practice of naturalists', Buckland complains, 'to follow Buffon in misrepresenting the Sloths as the most imperfectly constructed among all the members of the ani-mal kingdom, as creatures incapable of enjoyment, and formed only for misery.'[17] We can see Buckland's dilemma here, but he deftly works his way through the criticisms of sloths, especially the *Megatherium*, with almost Darwinian admiration – albeit with a Creationist flourish at the end – for the creature he calls the 'Leviathan of the Pampas'. 'His entire frame', Buckland explains about the *Megatherium*,

> was an apparatus of colossal mechanism, adapted exactly to do the work it had to do; strong and ponderous in proportion as this work was heavy, and calculated to be the vehicle of life and enjoyment to a gigantic race of quadrupeds, which, though they have ceased to be counted among the living inhabitants of our planet, have, in their fossil bones, left behind them imperishable monuments of the consummate skill with which they were constructed. Each limb and fragment of a limb forming co-ordinate parts of a well-adjusted and perfect whole; and through all their deviations from the form and proportion of the limbs of other quadrupeds, affording fresh proofs of the infinitely varied, and inexhaustible contrivances of Creative Wisdom.[18]

As historian of science Gowan Dawson has observed (drawing on the revisionist views of both Buckland and Owen), 'What had once seemed merely incongruous, ungainly, and awkward, was now instead revealed as an instance of harmonious and perfectly integrated design.'[19]

Whether sloths were considered odd and ungainly, or harmonious and perfectly designed, the fascination with giant ground sloths did not abate. The discovery of fossil sloths continued not only in South and Central America, but in North America, albeit a little more slowly. As G. Wayne Clough, the former secretary of the Smithsonian, has noted, a jawbone from a *Megatherium*, which 'originally had been donated by Dr James Proctor Screven in 1842 to an organization called the National Institute for the Promotion of Science in Washington, DC', eventually found its way into the Smithsonian collections in the 1850s.[20] Museums, particularly in the nineteenth century, were more about spectacle than education, and it simply made sense to include a giant mammal as a featured display in the collection. And so, as fossils emerged, including Darwin's *Mylodon* and *Scelidotherium* (complete with puzzling bony plates or osteoderms), the fascination with giant ground sloths intensified and was sustained throughout the nineteenth century.

Richard Owen, it is worth recalling, was perhaps among Britain's most celebrated and feared scientists in the period prior to Darwin, and his work in comparative anatomy and palaeontology has had an indelible imprint on these disciplines. While on the *Beagle*, Darwin sent many specimens directly to Owen, including fossils of giant ground sloths. As a result, Owen produced a series of memorable works, beginning in 1842, on the Mylodonts and the Megatheriid, and in many ways, as Frederick Waddy's caricature of Owen on his 'hobbyhorse' suggests, he was always connected with Megatheriids. One of the more interesting of Owens's observations was his conclusion that certain 'longitudinal fractures' that he noted on a particular *Mylodon* skull could not have been caused by a predator but were the result of either a tree limb or an entire tree falling on its head. As Owen's grandson and biographer explains:

> The probability was, then, that his habit was to uproot trees for the purpose of feeding upon their leaves, and once, when so doing, the tree must have fallen with a crash upon his skull, before he had time to move his huge carcass out of the way, and that this fracture had apparently no sooner healed than the same thing had happened again.[21]

The ludicrous idea of repeatedly fracturing one's skull while in the process of eating and having so huge a 'carcass' as not to be able to get out of the way did little to enhance the already-limited appreciation of sloths as intelligent beings.

SLOTHS IN VICTORIAN LITERATURE

It was surely Owens's description of the *Mylodon* that prompted naturalist, cleric and novelist Charles Kingsley (1819–1875) to depict the creature in the well-known evolutionary dream at the heart of the novel *Alton Locke* (1850). In one of his many progressive transformations, what he calls a 'strange jumble' of a dream, Alton finds that he has become a ground sloth, or in his own words,

a vast sleepy mass, my elephantine limbs and yard-long talons contrasting strangely with the little meek rabbit's head, furnished with a poor dozen of clumsy grinders, and

Richard Owen riding a fossil *Megatherium*, engraving by Frederick Waddy, 1873.

129

a very small kernel of brains, whose highest consciousness
was the enjoyment of muscular strength.[22]

He takes 'delight in tearing up trees', but he also explains, no doubt
taking his cue from Richard Owen, 'I had fractured my own skull
three or four times already.' Ultimately, Alton's dream-*Mylodon*
saves his cousin from a falling tree limb, only to have his back
broken and die 'in mortal agony'. It is, admittedly, a melodramatic
moment, but it served, for Kingsley, to mark that transition from
primitive beasts to a higher order of life.

What is important here is less the evolutionary allegory in the
dream than the fact that Kingsley can draw on the giant ground
sloth as a figure that all of his readers would recognize and could
place, as a consequence of Owen's work, into some historical context. If sloths are a twenty-first-century meme to us, the giant ground
sloth held a similar position in the minds of the Victorians, both
young and old. As early as 1848, the satiric magazine *Punch* proposed 'New Toys' for children including a rocking horse based on
the skeleton of a ground sloth. 'The Nimrods of the nursery', the
author writes, 'will ride nothing but a Megatherium.'

In 1854 the newly relocated Crystal Palace (which was moved
from Hyde Park to Sydenham) displayed 33 concrete replicas of
prehistoric creatures (consisting mostly of dinosaur-like creatures)
created by the artist Benjamin Waterhouse Hawkins. The exhibit,
which remains on display to this day, included an imposing sculp-
ture of a *Megatherium* standing impressively in an upright position.
Megatheriids appeared relatively frequently, even if only in passing references, in popular culture during the Victorian period.
Henry Morley, who worked with Dickens on the magazine *Household Words*, offered a view of 'Monsieur the Mylodon' consuming
a tree (and barely escaping a cracked skull) in his essay 'Our
Phantom Ship on an Antediluvian Cruise', which Victoria Coules

of the University of Bristol describes as one of the first examples of 'Palaeofiction'.[23]

William Makepeace Thackeray situates his character Pendennis in the fictional 'Megatherium Club' in *The Newcomes* (1855), perhaps inspiring William Stimpson, a zoologist at the Smithsonian, to form his own 'Megatherium Club' in 1857. Thackeray continues elsewhere in *The Newcomes* (1854–5), with his typical wit, to compare the careful assemblage of a literary production by the novelist to the work of palaeontologists. Thus, 'Professor Owen or Professor [Louis] Agassiz takes a fragment of a bone, and builds an enormous forgotten monster out of it, wallowing in primeval quagmires, tearing down leaves and branches of plants that flourished thousands of years ago.'[24] It is in this way, Thackeray explains, that the novelist can produce 'a more important animal, the megatherium of his history'.

Thackeray's invocation of *Megatherium* as a metaphor for the creation of fiction is a clear indication of a public awareness and even interest in these exotic fossils and their living descendants.

The popularity of giant sloths is made clear in this depiction of a child riding a sloth as a 'new' rocking *Megatherium*, from *Punch; or, The London Charivari* (1848).

One facet of that growing awareness was the emergent interest in the South American continent, not only because it appealed to Western naturalists, but because of its promise of wealth to colonial adventurists, not unlike the silver-mining Gould family in Joseph Conrad's *Nostromo* (1904). Nevertheless, the natural history of the South American continent was magnetic in its own right. Among the great nineteenth-century explorers who visited South America were the eccentric Charles Waterton (1782–1865) and the remarkably dedicated Henry Walter Bates (1825–1892). Waterton, to his credit, adopts a very pragmatic view of the sloth, which, as he observes in his *Wanderings in South America* (1825), without slipping into received ideas about sloths, both responds to changes in the weather and follows the need to forage. The sloth, Waterton notes, 'travels at a good round pace; and were you to see him pass from tree to tree, as I have done, you would never think of calling him a sloth'. Waterton proceeds to chastise less dedicated naturalists who have mischaracterized the creature. 'The different histories we have of this quadruped', Waterton points out, 'are erroneous on two accounts':

> first, that the writers of them deterred by difficulties and local annoyances, have not paid sufficient attention to him in his native haunts; and secondly, they have described him in a situation in which he was never intended by nature to cut a figure; I mean on the ground. The sloth is as much at a loss to proceed on his journey upon a smooth and level floor, as a man would be who had to walk a mile in stilts upon a line of feather beds.[25]

Waterton, for all of his eccentricities (and they were many), was a very close, thoughtful and perhaps, one might say, sympathetic observer. He notes, with a tinge of conventional Victorian thought, that 'the sloth is doomed to spend his whole life in the trees; and what is more extraordinary, not upon the branches, like the squirrel and the monkey, but under them.' But Waterton then, with the ecological sensibility that marked much of his thought, astutely recognizes that 'to enable him [the sloth] to do this [live suspended from branches], he must have a very different formation from that of any other known quadruped.' His conclusion, rather than treat the sloth as a freakish anomaly, a 'bungled' production of nature, anticipates what we would now call the adaptive fitness of the animal. 'Hence', Waterton observes, combining his zoological knowledge with his Catholic piety,

> his seemingly bungled conformation is at once accounted for; and in lieu of the sloth leading a painful life, and entailing a melancholy and miserable existence on its progeny, it is but fair to surmise that it just enjoys life as much as any other animal, and that its extraordinary formation

Sloths, as Charles Waterton noted, were well adapted to their unconventional lifestyle.

and singular habits are but further proofs to engage us to admire the wonderful works of Omnipotence.[26]

The extraordinary popularity of Waterton's *Wanderings* attracted many young readers, including Darwin, Alfred Russel Wallace and Henry Walter Bates, all three of whom eventually made their way to South America. Wallace and Bates, who had become friends in their youth, decided to explore the Amazon rainforest with an eye to selling specimens to support their journey. Bates's legacy includes the concept of Batesian mimicry, which occurs when harmless species, like flies, mimic the traits of aggressive or toxic species, such as wasps and monarch butterflies, that predators tend to avoid. Bates devoted eleven years of his life to studying the Amazon and eventually published his own account of his findings in *The Naturalist on the River Amazons* [*sic*] (1863). Impressed with the sloth's ability to secure its hold from one branch to the next before advancing, and then to maintain that grip, Bates describes shooting the animal, noting that it remained in place until it must have 'dropped on the relaxation of the muscles a few hours after death'. To be sure, nineteenth-century practices of natural history were radically different from current techniques, so it is not surprising that after Bates and Wallace noted with delight that sloths were competent swimmers, Bates dispassionately recorded that 'our men caught the beast, cooked, and ate him.'[27]

We can find a more empathetic view of the sloth in Elizabeth Agassiz's *A Journey in Brazil* (1868), in which she describes a sloth, one of many 'pets' on board the boat that is taking her up the Amazon. 'I am never tired of watching him,' she writes:

he looks so deliciously lazy. His head sunk in his arms, his whole attitude lax and indifferent, he seems to ask only for rest. If you push him, or if, as often happens, a passer-by gives him a smart tap to arouse him, he lifts his head and drops his arms so slowly, so deliberately, that they hardly seem to move, raises his heavy lids and lets his large eyes rest upon your face for a moment with appealing, hopeless indolence.[28]

It is probable that the sloth 'pet' that was brought on board by a crew member was a maned sloth (*Bradypus torquatus*), but Agassiz doesn't identify the species. Alas, given the average life span of captive Bradypodids (even as pets on the Amazon), it is unlikely that this animal, the source of her delight, had a particularly bright future.

Beyond Agassiz's reflection of interacting with a sloth, there were no lengthy studies of sloths in general for a good long while. Finally, a comprehensive overview of sloths, perhaps the most detailed to date, was undertaken by William Beebe (1877–1962) in his essay 'The Three-Toed Sloth', which was published in 1926 in *Zoologica*, the journal of the New York Zoological Society. Beebe, a thorough naturalist and adventurous explorer, studied sloths while working in what is now Guyana. Beebe begins his lengthy essay with a tongue-in-cheek assessment of the sloth, suggesting that they 'have no right to be living on this earth, but they would be fitting inhabitants of Mars, whose year is over six hundred days long'.[29] The remainder of the text, however, while not always accurate by contemporary standards, is serious, thorough and detailed.

SLOTHS IN THE AGE OF MASS MEDIA

Television

Beebe's work aside, interest in sloths in the twentieth century seemed to slow down, as it were. In isolated instances, interest was strong, but the resurgence of interest in arboreal sloths is a very recent phenomenon and can be attributed to natural history shows on television in the 1980s and '90s. In 2006 the American Public Broadcasting Service (PBS) launched an animated nature show with the title *It's a Big Big World*, which featured a character called Snook the Sloth as its host. Snook – who, despite being a pale-throated three-toed sloth, actually has a thumb and a big red nose – was voiced by the puppeteer Peter Linz. Although Snook was a genial and very laid-back character (one might even say 'Dude-like'), many children mistook him for a bear, and some were afraid of him. Only 47 episodes of the show were produced, although reruns continued until 2010; in subsequent years, Snook's creator, Mitchell Kriegman, toured with a giant costumed Snook to support environmental events.

Music

Sloths, although anything but musical themselves, were the inspiration for a garage rock band called The Sloths that initially formed in Los Angeles in the 1960s and achieved local popularity but little renown. Undaunted by the passage of time, however, the band performed periodically and gradually – living up to its name – and (at long last) found its way to releasing the album *Back from the Grave* in 2015, with a cartoonishly gruesome cover and array of songs, which, though sloth-free, have a compelling retro sound.

By contrast, Trey Anastasio's band Phish produced over ten albums over the course of twenty years. In one of their songs, we actually hear – in the first person – from a sloth. Anastasio composed 'The Sloth' for his college senior thesis as part of a concept album called *The Man Who Stepped into Yesterday* (1987), and even though that album was never released, 'The Sloth' has been performed many times by Anastasio with Phish. The lyrics are meant to capture the spirit of an unpleasant and possibly disturbing

Cover art for the
Los Angeles-based
group The Sloths
depicting zombie
sloths.

sloth-like character that figures in a much larger narrative created by Anastasio:

They call me the sloth
(Way down in the ghetto)
Italian Spaghetti
(Singing falsetto)
Sleeping all day
(Rip Van Winklin')
Spend my nights in bars
(Glasses tinklin')

In stark contrast to Anastasio's sloth is perhaps the most charming depiction of sloths in prose and verse. 'The Sloth', penned by the entertainers Michael Flanders and Donald Swann in the 1960s, exudes humour, charm and a certain level of enchantment. The introspective lyrics, in the voice of the sloth itself (Michael Flanders), have everything to do with lingering Victorian values, but at the same time, they touch on the tension we all feel between feelings of brazen aspirations and pragmatic concessions. Alive with clever wordplay, the song takes into account both British and American pronunciations of 'sloth' and is well worth quoting in its entirety:

A Bradypus or Sloth am I,
I live a life of ease,
Contented not to do or die
But idle as I please;
I have three toes on either foot
Or half-a-doz on both;
With leaves and fruits and shoots to eat
How sweet to be a Sloth.

The world is such a cheerful place
When viewed from upside down;
It makes a rise of every fall,
A smile of every frown;
I watch the fleeting flutter by
Of butterfly or moth
And think of all the things I'd try
If I were not a Sloth.

I could climb the very highest Himalayas,
Be among the greatest ever tennis players,
Always win at chess or marry a princess or
Study hard and be an eminent professor,
I could be a millionaire, play the clarinet,
Travel everywhere,
Learn to cook, catch a crook,
Win a war, then write a book about it,
I could paint a Mona Lisa,
I could be another Caesar,
Compose an oratorio that was sublime:
The door's not shut
On my genius but
I just don't have the time!

For days and days among the trees
I sleep and dream and doze,
Just gently swaying in the breeze
Suspended by my toes;
While eager beavers overhead
Rush through the undergrowth,
I watch the clouds beneath my feet—
How sweet to be a Sloth.

The wonderfully inert world view captured by Flanders and Swann is evocative because the sloth is speaking for itself in the first person (or 'first sloth'). Even though this was essentially written as a song for children, it seems to distil the essence of the sloth (albeit anthropomorphically), easily capturing adult imagination.

Cinema and social media

The years 2013 and 2014 saw the production of two sloth documentaries. The earlier documentary, produced as part of the *Animal Planet* series and titled *Meet the Sloths*, ostensibly followed the operations and animals at the Sloth Sanctuary in Limón province, Costa Rica. The central human figures in the documentary are Judy Avey-Arroyo, who transformed her hotel into a sanctuary after receiving an orphaned baby sloth in 1992, and the zoologist Rebecca Cliffe, who was, at the time, conducting research for her doctorate and is also the founder of the Sloth Conservation Foundation (SloCo), launched in 2017. The Sloth Sanctuary has attracted a degree of controversy with respect to the number of animals it has rescued and that remain captive in the sanctuary. Holding rescued animals and ultimately rehabilitating them is both complex and difficult, and consequently the controversy has not abated. Another central character of the *Animal Planet* series was Buttercup the Sloth, perhaps the oldest living *Bradypus* in captivity, who died in 2019 at the estimated age of 27. The two episodes, though highly sentimentalized (in the spirit of many animal documentaries), provide an interesting overview of the care, feeding and behaviour of both *Bradypus* and *Choloepus*. In 2014 PBS released *A Sloth Named Velcro*, directed by journalist Ana Salceda, which addresses sloth biology and behaviour while focusing on Salceda's own relationship with an orphaned *Choloepus* sloth that she eventually releases back into the wild.

The documentary also explores in serious detail the environmental crises faced by sloths in the wild.

So many other depictions of sloths involve human–animal interactions that are more likely than not to make the sloth itself an object of consumption. Perhaps the most popular sloth moment on YouTube was Kristen Bell's response to receiving a 'sloth experience' on her 31st birthday.[30] In the video, which has had up to 30 million views, Bell, who is clearly passionate about sloths, is overcome with excitement about having a sloth 'visit' her party. Bell's experience was undoubtedly engaging and fulfilling for her, but it probably wasn't the healthiest experience for the sloth, particularly because three-toed sloths are not especially social.

Sloth tourism and captivity

There are plenty of sloth encounters available, mostly on site at zoos or animal parks, at costs ranging from $100 (£75) to $450 (£325), and most of these sites use two-toed sloths as part of the experience. The advertising often betrays exactly what's wrong with these encounters, which have less to do with understanding the animal than achieving personal or 'spiritual' gratification. One site, invoking the familiar phrase 'Wanna Get Slothed', offers 'an intimate experience that is fun for the whole family and will give you a whole new perspective – teaching us to slow down, be grateful and enjoy each moment'.[31] This kind of rhetoric not only undermines a clear understanding of how animals, sloths or otherwise, exist in the world, but distorts our view of them by rendering them as creatures made to represent human needs. Surely there are better, and less traumatic, ways for interested individuals to observe and learn about the animals they claim to cherish so deeply.

There are at least forty locations in the United States that have sloths on display or that actually allow patrons to interact with the animals. The number of captive sloths in Europe seems to be much lower, but accurate counts are not available. In the UK, Drusillas Park in Alfriston, East Sussex, is one of at least seven small zoos in which you can book time for a 'close encounter' with one of their resident sloths. The result of this interaction, according to the Drusillas website, is an 'amazing one-to-one experience lasting between 30 and 40 minutes which will leave you grinning from ear to ear'. It stands to reason that the encounter may not be nearly as rewarding for the sloth.The price of a typical sloth experience varies between £59 and £95 for a single individual, but certain deluxe packages may cost as much as £200. Equally disturbing is the impact of what is often called selfie tourism. Recently, organizations such as World Animal Protection have studied human–sloth interactions, and their preliminary findings about the way that humans treat sloths – often inadvertently – when handling them are not favourable: they are more than likely 'compromising [the sloths'] welfare'. When handled by tourists, the sloths display unusual behaviours that 'may be indicators of fear, stress and anxiety', which are not justified by their treatment as photo props.[32] No less disturbing are the organizations, often sheltered under not-for-profit umbrellas, that offer to rent sloths for special occasions. The fee, for at least one company in the USA, is $400 for the initial 'rental' and $350 for every additional hour.

Another intrusive form of interaction with animals are drones, owned and operated by tourists and possibly even documentarians. On 12 May 2021, for example, a drone crashed into a colony of elegant terns (*Thalasseus elegans*, a species of sea bird) at the Bolsa Chica Ecological Reserve south of Los Angeles, leading to the abandonment of over 1,500 nests and a serious disruption of the population.[33] Even if less dramatic circumstances result, unregulated drones flying through tree canopies in search of sloths will surely affect the behaviour of the animals and their offspring in ways that we can't and, quite frankly shouldn't, have to predict. Many wildlife parks are considering how to limit or possibly even ban the use of drones given how intrusive and potentially dangerous they may be in terms of native species.

Sloth encounters, whether through travel or zoo-like programmes, are very much a modern phenomenon, but they are still relatively rare experiences for both adults and children. Needless to say, there are countless YouTube videos of sloths, showing them feeding, climbing, swimming and even copulating, but mostly presenting them as cute and cuddly.

Books

Beyond videos, sloths, often for better but sometimes for worse, have a prominent place in children's literature. A number of non-fiction books by established authors, such as Rebecca Cliffe (*Sloths: Life in the Slow Lane*), Lucy Cooke (*A Little Book of Sloth*), Bonnie Bader (*Slow, Slow Sloths*), Sam Trull (*Slothlove*) and Josh Gregory (*Sloths*), are competent, accurate and informative in presenting details about the biology and environment of sloths, yet they are also amusing and engaging. Among the best and most creative picture books about sloths is Eric Carle's '*Slowly, Slowly, Slowly*', *Said the Sloth* (2002), which is introduced by Jane Goodall, and attempts to present the lifestyle of the sloth as a natural adaptation to its habitat. *Wake Up, Sloth!* (2011), the collaborative effort of Anouck Boisrobert and Louis Rigaud, is an engagingly serious pop-up book that details the habit destruction in the forest surrounding a sloth. Ultimately, the authors suggest that the relentless decline of tropical habitats is having a severe impact on sloth populations; nevertheless, they also offer hope that with increased care and attention to the forests, the sloths will be able to make a comeback.

Fictional works can be more problematic when it comes to presenting the sloth without resorting to clichés and formulaic platitudes about lazy sloths. *Julie's Secret Sloth*, though published in 1953, is still a troubling case in point about a young girl who acquires a sloth from a failing menagerie and then attempts to hide it from friends and family.[34] *Sparky* (2014), by the novelist Jenny Offill and illustrated by the talented Chris Appelhans, though more recent than *Secret Sloth*, seems no more sensitive or humanely aware than its predecessor. *Sparky* begins its narrative with a little girl who is told she can only have a pet that 'doesn't need to be walked, bathed, or fed'. As a consequence, she actually orders a sloth by 'express mail', which, though legal in some locations, is frowned upon and seems out of place in a book for children.

Not only are many sloths endangered, but they are very difficult to keep in captivity, much less as pets for children; most will

Cover for Eric Carle, *'Slowly, Slowly, Slowly,'* *Said the Sloth* (2022).

Sylvaine Peyrols' depiction of a sloth, the caption reading: 'With its round face, the sloth always seems to be smiling.'

The seemingly gentle facial expression of a three-toed sloth.

eventually die in a matter of weeks or months. There is an illicit, yet robust, trade in exotic animals, even in countries such as Colombia, where the practice is illegal. It probably goes without saying that sloth 'vendors' are sketchy at best and have little concern either for the welfare of the animal or the suitability of the prospective owners. Sloths are sold for anywhere between $125 and $10,000. Sparky, to return to our story, turns out (unsurprisingly) to be laconic and disappointing; but more important for readers of all ages is the travesty inherent in commercializing an exotic animal for the sake of the pet trade. Other books seem to want to capitalize on the popularity of sloths, but aside from being represented as slow and frequently foolish, the sloth characters have little depth and less detail.

One of the more striking children's books about sloths is, thus far, only available in French. *Le Paresseux*, illustrated by Sylvaine Peyrols, is book number 78 in the Gallimard series *Mes premières découvertes* (My First Discoveries), which focuses predominantly on animals. The text is simple but informative and accurate, even to the point of explaining the sloth bathroom habits: 'He hollows out a small hole, where he deposits several very dry turds.'[35] Peyrols's images are, for the most part, lifelike, although the 'smiling' sloth exudes a little extra imaginative charm.

Animated films

Sloths have achieved prominence in several animated movies in the last two decades, including the *Ice Age* films (2002, 2006, 2009, 2012, 2016), *The Croods* films (2013, 2020) and, most recently, *Zootopia* (2016). *Ice Age* has been the most successful franchise in part because of John Leguizamo's portrayal of Sid the Sloth, a comedic character that is ostensibly a *Megalonyx*, although

Sid the Sloth, one of the central characters of the *Ice Age* series of animated films (2002–).

147

that's never made clear in the movies. What's more, Sid is a little short in stature to be a full-fledged ground sloth, but that's only one inaccuracy in a series of films created for childhood entertainment. Still, one might want to wish away the conflation between ice age mammals and the reptiles, such as *Tyrannosaurus*, of the late Cretaceous that form part of the premise of *Ice Age: Dawn of the Dinosaurs* (2009). While these movies are clearly not educational, it is unfortunate that the 'ice age' theme now suggests that large mammals and dinosaurs lived, as it were, mammalian cheek by reptilian jowl.

Giant ground sloths make brief appearances in the television shows *The Simpsons* and *Futurama*, two animated series developed by Matt Groening. In both cases, the ground sloths are absurdly slow; in 'Super Franchise Me' (2014), caveman Homer Simpson, slothful in his own right, pursues a giant sloth and eventually subdues it, while in a 2012 episode of *Futurama*, 'Fun on a Bun', a giant sloth is released as an attack animal but strikes so slowly that it's clearly no threat at all. The episodes offer a knowing and ironic view of sloths, ancient and modern, with an eye to a clever and perceptive viewership.

A more violent version of a ground sloth variant does appear as the character Megasloth in Fallout 76, a role-playing game by Bethesda Softworks. According to one description of the avatar, 'megasloths are three-toed sloths mutated by radiation. Standing much taller than the average human, the sloths are covered with filthy and shaggy fur with clusters of mushrooms growing in the fur on their back[s].' From all appearances, these sloths have every conventional attribute of a video-game threat but few qualities of either tree sloths or, for that matter, ground sloths.

The Croods, an animated franchise by DreamWorks Animation, offers a reverse anachronism given that the Crood family unit are ostensibly and implausibly from the Pliocene (5.3 to 2.6 million years ago). Their pet sloths (who are disguised as both a belt and a sash) are versions of more contemporary Bradypodids, which are unlikely to have assumed their current form in that period. Once again, it would be absurd to give any scientific credence to movies like *The Croods*, and so pointing out these anachronisms seems overly fussy. And yet if there is enthusiasm among young viewers for stories in prehistory, and if we credit their knowledge and intelligence, it might a good idea to be less blasé about zoological history.

The film *Zootopia*, produced by Walt Disney Animation Studios, abandons history and even a natural setting for its animals in favour of a cosmopolitan and anthropomorphic world. The film was a remarkable success and became the fourth-highest-grossing film of 2016. Drawing on clever dialogue and familiar social situations to appeal to both children and adults, it received kudos for addressing issues of tolerance, prejudice and equity. The animals in the movies are thus allegorical, following a long tradition in

Flash the sloth from the Disney film *Zootopia* (2016).

literature and film from the medieval *Reynard the Fox* (*c.* 1250) to Orwell's *Animal Farm* (1945). Still (and this may be a bit unreasonable), the movie also represents an idealized society of animals living together in abundance, even though many of the represented animals are endangered and threatened. Nevertheless, one of the break-out characters from the movie is Flash, the sloth who works with other sloths including Priscilla (in a neighbouring cubicle) at the Department of Mammal Vehicles (that is, the DMV, the equivalent to the DVLA in the UK). A three-toed sloth (albeit with a thumb), Flash and his colleagues are a wonderful parody of the frustration nearly everyone has in dealing with the bureaucracy of obtaining a licence to drive, number plates or both. Next to Flash is a coffee mug that reads: 'You Want It When?', which, of course, has become a very popular merchandise item generated by the movie.

Sloths as memes
The sloth has also become an Internet meme because of its anthropomorphic cuteness and the impression that it is both infantile and helpless (repeating, in an updated way, eighteenth-century attitudes). The impression that the sloth is laid back renders it an emblematic antidote to the frenetic pace of Western culture. *Time* magazine wrote in 2013 – the year that *Animal Planet* showed a documentary by Lucy Cooke called *Meet the Sloths* –

> The Google trends graph of searches for the term 'sloth' was once a series of modest peaks and valleys, reflecting the fluctuating interest in the animal (or deadly sin). Then everything changed. Interest now is historically high; earlier this week, BuzzFeed published its epic list of the 25 Greatest Sloths the Internet Has Ever Seen.[36]

The ubiquitous world of Pokémon saw the invention of its first
sloth-like creature, Slakoth, in 2003–4; it evolves into two later
forms, Vigoroth and Slaking, both sloth-like in their own right.
While Slakoth appears to be a two-toed sloth, it has some of the
colouration of a three-toed sloth; nevertheless, it is said to sleep
twenty hours per day, which simply suggests that it is slothful.
The quizzical look and flattened posture of Slakoth are stereo-
typical of the perception of fully inactive sloths, and no doubt that
helps in its appeal to children.

Graphic novels
The characterization of individuals as sloths also continues to
have a dark side, as the renowned author/illustrator Gilbert
Hernandez's work *Sloth* suggests. A graphic novel intended for
'mature readers', Hernandez's book traces the lives of three

teenagers who all belong to a band called Sloth.[37] More important is that each character, beginning with Miguel, is bullied and called a sloth as a consequence of slipping into a coma. Two of the other characters experience comas or coma-like states over the course of the novel; each coma represents a form of social withdrawal, perhaps more characteristic of teenage angst than sloth behaviour.

Among the more recent popularizations of sloths is *Slothilda*, a character created by cartoonist Dante Fabiero. The character, featured in cartoons and books, reminds us, according to Fabiero,

Slothilda, a playful character from a book series by Dante Fabiero.

'to feel unashamed about our slothy tendencies and emphasizes the importance of celebrating our authentic selves'.

A recent series of books by Canadian artist and animator Graham Annable follows the adventures of Peter and Ernesto, two loosely conceived three-toed sloths, the former a homebody and the latter an adventurer. With respect to studies of sloth behaviour, the very idea of sloths experiencing friendship runs counter to virtually everything we know about their habits, but these are fantastic tales written for children. The books are engaging for young readers and progressively include more native species from Central and South America, thus introducing something of a broader ecological context.

Of course, sloths are neither infantile nor helpless, nor even lazy, and the impression of them as caricatures of living creatures undermines a serious understanding of their zoological identities and significance. That said, sloth iconography is markedly ubiquitous. Plush sloths abound, as do T-shirts, mugs, tea-infusers, jewellery, keychains, calendars and even snow-globes. A search for 'sloth' on Etsy will satisfy anyone obsessed with sloths, offering anything from sloth pots (for succulent plants) to a Harry Potter-style keychain labelled Hairy Slother. Of course, Etsy, as many shoppers know, is an endless repository of oddly themed merchandise, but there does seem to be a particular effusion of sloth-related items. Taking nothing away from Etsy, it is true that there are many sites in which sloth-related purchases will benefit environmental and rehabilitative efforts (in some way). In any event, it seems likely that the trend will continue, although one hopes that the sloth emerges from this merchandising wave as something more than a cute rendition of ironic malaise or fatigue.

In the face – quite literally – of cute sloth memes (and merchandise), James Tynion IV, Eryk Donovan and Adam Guzowski

produced the graphic novel *Memetic* (2015), in which a 'weaponized' sloth meme (with an odd thumb) is universally distributed in all media by a mysterious (perhaps alien) source. The captivating meme, which is smiling and deceptively cute, paradoxically mesmerizes every human on earth to become mindlessly violent, and this inexorably leads to total annihilation in roughly 72 hours. It is a bleak topic for an illustrated work, but it does conform to a common genre of apocalyptic endings in graphic novels. The work holds its own as a satire of a smartphone-addicted culture that is itself enthralled by trendy, and more often than not vacuous, memes.

The sloth meme that paralyses all onlookers, from James Tynion IV's graphic novel *Memetic* (2015), with art by Eryk Donovan and colouring by Adam Guzowski.

6 Sloths, Conservation and Eco-Awareness

If we provide nature with the minimum it will make
the most of it and will reward us with blossoming life.
Sloth researcher Antonio Rossano Mendes Pontes[1]

The ecological niche occupied by every single animal is linked, often in imperceptible ways, to myriad other facets of the environment. It is difficult to situate sloths in the larger ecological frame of the rainforest environment since so many studies are intensely species-focused. Nevertheless, we can say that in some areas of South America, sloths and howler monkeys comprise the most substantial portion (roughly 50 per cent) of the mammalian biomass.[2] This is hardly trivial given that they must inevitably provide a sizeable percentage of the prey consumed by eagles, jaguars, ocelots and anacondas. That's not to say that there aren't other mammalian prey species; in fact, there is an impressive diversity of mammalian fauna in Central and South America, ranging from a wide array of opossums, anteaters and tamanduas, agoutis, capybaras and other New-World monkeys, as well as squirrels and other rodents. In addition, as bird enthusiasts know, the avian fauna of Central and South America is incredible, as is the insect life. Sloths, we must remember (despite the all-encompassing title of this book), should be set against a backdrop of astonishing and, it must be said, vulnerable, biodiversity.

To return our focus to sloths, however, there is an impressive biodiversity simply associated with the animals themselves. We know that all animals support, internally, a micro-biome that involves a full complement of bacteria that assist every species with

Grogu, an
orphaned sloth,
Costa Rica.

Sloths are known to traverse flat terrain, if necessary, but increased clear-cutting and the building of roads has made land travel even more treacherous.

digestion, immunity and eliminating toxins. Other relationsips are 'mutualistic', a system in which two species live in tandem, and each benefits from the partnership. These 'partners' may range worldwide from 'cleaning wrasses' and groupers to ox-peckers and rhinos. The sloths enjoy a symbiotic relationship, as we have seen, with snout moths and a variety of insects, to say nothing of algae resident on their fur. Brown jays (*Psilorhinus morio*) have recently been observed snacking on the fur of the Bradypodids, but whether they are simply getting a free meal of providing some advantage to the sloth is not clear. Certainly, if they are eating ticks, mites and blood-sucking insects (triatomine bugs) there is a benefit to the sloth itself.[3] Sloth dung, in addition to supporting moth larvae, may include seeds that have some nutritional value for scarab beetles, although as one author observes, their scat 'is dry, hard, and pelletized'.

We will only begin to understand the full ecological profile of sloths, whether physiologically, behaviourally or environmentally, by further observational and unobtrusive studies. But our increased understanding and appreciation of sloths is also contingent on their survival, which even for the most populous species is by no means certain. Agricultural clear-cutting is a serious threat to sloths, as for many other animals and plants. The intrusion of roads (very dangerous when crossed on the ground) and powerlines (which often lead to electrocution) are serious threats to sloths. In some areas, rope bridges provide some relief (albeit very limited) by allowing sloths to move from one area to the next without the threat of being run over or electrocuted. Often these roads and power lines serve the growing populations where large-scale agriculture continues to consume acres of rainforest. Large monoculture plantations – of palm oil and sugar cane for example – not only fragment the landscape but introduce pesticides well beyond their borders, and a number of alarming deformities have been observed among sloths in recent years. Wildlife in general depends on habitable 'corridors',

A sloth using one of many rope bridges to cross over areas that have been clear cut or have electrical lines.

where animals can find sustenance, migrate and find refuge. Los Angeles, for example, is creating a long-overdue wildlife corridor over California's Highway 101, linking disparate parts of the Santa Monica Mountains.[4] As arboreal animals, sloths in particular need standing forests of trees in order to be able to continue normal foraging and mating behaviours. In spite of moving very slowly, sloths do move regularly from tree to tree and require adequate space to survive and thrive. That movement also increases genetic diversity, as opposed to the prospect of inbreeding when populations of sloths are essentially restricted to small areas.

Survival is most dire for the pygmy three-toed sloths (*Bradypus pygmaeus*), which is listed as 'Critically Endangered' by the IUCN; the maned sloth (*Bradypus torquatus*) in Brazil, which has been designated as 'Vulnerable', requires equally serious attention. The *New York Times* has noted that despite promises to protect the environments, Brazilian president Jair Bolsonaro (2019–22), 'has worked relentlessly and unapologetically to roll back enforcement

Warning signs like these are becoming very common wherever sloth populations are found.

of Brazil's once-strict environmental protections'.[5] The legitimate
hope, bolstered by the recent election of Luiz Inácio Lula da Silva
(Lula) in Brazil, is that political leaders will step up to honour
their own environmental responsibility and that of their country.
And while many political leaders remain environmentally regres-
sive, voters, activists, environmentalists and even politicians must
continue to be both vigilant and relentless.

Choloepus and
Bradypus in two
perspectives.

In Panama, the lack of sufficient legal protection for pygmy
three-toed sloths and their habitat has caused a sharp decrease in
their population, a situation that is exacerbated by destruction
and fragmentation of their habitat, some exploitation for food and
the presence of feral cats. While their native island is essentially
uninhabited, seasonal visitors, including fishermen, lobster divers
and occasional Indigenous poachers have been known to hunt the
sloths. The recent fame of sloths has also made them a target for
capture and export for public display. An attempt to export eight
pygmy sloths by the Dallas World Aquarium in 2013 was thwarted
after local environmentalists, police, Indigenous groups and Panamanian authorities negotiated the surrender of the sloths from
aquarium representatives.[6] Still, at least two of the captured sloths
died prior to being released back into the wild. Nisha Owen, of

Bradypus variegatus

the Zoological Society of London is working with veterinarian Diorene Smith on an elaborate project to conserve the population (of pygmy sloths), along with their environment, in what is an ambitious but admirable undertaking.[7]

The sloth's status as a meme for cuteness and casual indifference seems to be the dominant theme of most representations of the sloth in Western culture. To a great extent, that militates against a more nuanced understanding of the sloth as a well-adapted organism in a very complex environment. What makes that environment particularly complex, to repeat what the reader surely knows by now, is the extent to which it is threatened by deforestation, pollution and even eco-tourism. What's more, the role of the sloth within the ecosystem that sustains it is not as well understood as it should be. One can only hope that the many attractions of sloths and their wonderful appeal in social media will eventually result in a more comprehensive understanding of them as organisms rather than disembodied memes or living toys for human amusement.

Timeline of the Sloth

65–60 MYA	47 MYA	35 MYA	18 MYA
Xenarthrans, the parent group of sloths, armadillos and anteaters, emerge in the Western Hemisphere	*Pseudoglyptodon*, the oldest sloth, appears	Oldest Mylodontodid and Megatheriid sloths appear, marking a clear division between two-toed and three-toed sloths	Giant ground sloths emerge 

1175	1535	1555	1758
The word *slauðe*, or 'sloth', appears in Middle English	Gonzalo Fernández de Oviedo (1478–1557) refers to sloths as *periquitos ligeros* or 'swift parrots' 	André Thévet visits Brazil; he publishes *Singularitez de la France antarctique* in 1557, which provides perhaps the first Western illustration of a sloth	Linnaeus classifies the two-toed sloth, *Choloepus didactylus*, which becomes known as Linnaeus' two-toed sloth, and the three-toed sloth, *Bradypus tridactylus*

1834	1858	1861	1960s
Charles Darwin, voyaging on *HMS Beagle*, sends fossil ground sloths to London	The northern two-toed sloth *Choloepus hoffmanni* is named after the German naturalist Karl Hoffmann, who lived in Costa Rica	Richard Owen publishes *Memoir on the Megatherium* 	Marion P. McCrane of the Smithsonian's National Zoo successfully raises a two-toed sloth

3 MYA	25,000 YA	11,000 YA	345–399 CE
The formation of the Panama Isthmus allows northern sloth migration (the Great American Biotic Interchange)	The ice age pushes sloths back to southern regions	Giant ground sloths disappear	Evagrius Ponticus compiles a list of eight cardinal sins including 'sloth'

1795	1799	1804	1825
Juan Bautista Bru (1740–1799) assembles a giant sloth skeleton at the Royal Cabinet of Natural History, Madrid	Thomas Jefferson publishes 'A Memoir on the Discovery of Certain Bones of a Quadruped of the Clawed Kind in the Western Parts of Virginia'	Georges Cuvier publishes *Sur le Megatherium* (1804)	Charles Waterton vindicates sloths in his *Wanderings in South America*

1971	2001	2002	2016
M. Goffart publishes *Function and Form in the Sloth*	Pygmy three-toed sloths (*Bradypus pygmaeus*) identified as a new species	Sid the Sloth is a sensation in the *Ice Age* film franchise	Flash, the Department of Mammal Vehicles sloth, appears in *Zootopia*

References

PREFACE

1   Annette Aiello, 'Sloth Hair: Unanswered Questions', in
    The Evolution and Ecology of Armadillos, Sloths, and Vermilinguas,
    ed. G. Gene Montgomery (Washington, DC, 1995),
    pp. 213–18.
2   Susan Bronson, 'The Design of the Peter Redpath Museum
    at McGill University: The Genesis, Expression and Evolution
    of an Idea about Natural History', *Bulletin: Society for the
    Study of Architecture in Canada/Bulletin de la Société pour
    l'étude de l'architecture au Canada*, XVII/3 (September 1992),
    p. 69. The inscription is from Psalm 104:24 in the King
    James Version.
3   Robert Chambers, *Vestiges of the Natural History of Creation*
    (London, 1844), p. 130.
4   Henry Augustus Ward, *Catalogue of Casts of Fossils, from the
    Principal Museums of Europe and America, with Short Descriptions
    and Illustrations* (Rochester, NY, 1866), adj. to p. 10.
5   J. W. Dawson, *The Story of Earth and Man* (New York, 1873),
    p. 281.
6   Anon., 'Reception for the American Association for the
    Advancement of Science in the New Redpath Museum',
    Canadian Illustrated News, 2 September 1882, p. 153.

1 THE MODERN SLOTH

1   Jane Goodall, Preface to *'Slowly, Slowly, Slowly,' Said the Sloth*
    (New York, 2002), n.p.

2 Timothy J. Gaudin, 'Phylogenetic Relationships among Sloths (Mammalia, Xenarthra, Tardigrada): The Craniodental Evidence', *Zoological Journal of the Linnean Society*, CXL/2 (February 2004), pp. 255–305.
3 Kenny J. Travouillon, Western Australian Museum, personal communication, 6 January 2021.

2 ANATOMY AND PHYSIOLOGY

1 Georges Louis Leclerc, Comte de Buffon, *Natural History*, ed. James Smith Barr (London, 1792), vol. VIII, p. 290.
2 B. Urbani and C. Bosque, 'Feeding Ecology and Postural Behaviour of the Three-Toed Sloth (*Bradypus variegatus flaccidus*) in Northern Venezuela', *Mammalian Biology*, LXXII/6 (November 2007), pp. 321–9.
3 B. Voirin et al., 'Why Do Sloths Poop on the Ground?', in *Treetops at Risk*, ed. M. Lowman, S. Devy and T. Ganesh (New York, 2013), pp. 195–9.
4 Nadine Jacobson, 'Alarm Bradycardia in White-Tailed Deer Fawns (*Odocoileus virginianus*)', *Journal of Mammalogy*, LX/2 (May 1979), pp. 343–9.
5 R. N. Cliffe et al., 'Mitigating the Squash Effect: Sloths Breathe Easily Upside Down', *Biology Letters*, X/4 (April 2014), article ID 20140172.
6 Olivier Lambert, Giovanni Bianucci and Christian de Muizon, 'Macroraptorial Sperm Whales (Cetacea, Odontoceti, Physeteroidea) from the Miocene of Peru', *Zoological Journal of the Linnean Society*, CLXXIX/2 (February 2017), pp. 404–74.
7 Sarah Higginbotham et al., 'Sloth Hair as a Novel Source of Fungi with Potent Anti-Parasitic, Anti-Cancer and Anti-Bacterial Bioactivity', *PLOS ONE*, IX/1 (January 2014), 10.1371.

3 SLOTH SPECIES AND EVOLUTION

1 Charles Darwin, *The Correspondence of Charles Darwin*, vol. I: *1821–1836*, ed. David Kohn and F. Burkhardt (Cambridge, 1985), p. 276.

2   Raj Pant et al., 'Complex Body Size Trends in the Evolution of Sloths (Xenarthra: Pilosa)', *BMC Evolutionary Biology*, XIV (September 2014), art. no. 184.

3   Thomas Jefferson, 'A Memoir on the Discovery of Certain Bones of a Quadruped of the Clawed Kind in the Western Parts of Virginia', *Transactions of the American Philosophical Society*, IV (1799), pp. 246–60.

4   R. A. Fariña and R. E. Blanco, '*Megatherium*, the Stabber', *Proceedings of the Royal Society of London B*, CCLXIII/1377 (December 1996), pp. 1725–9.

5   Ibid., p. 1725.

6   R. A. Fariña, '*Megatherium*, the Hairless: Appearance of the Great Quaternary Sloths (Mammalia; Xenarthra)', *Ameghiniana*, XXXIX/2 (June 2002), pp. 241–4.

7   Anon., 'The Quadrupeds' Pic-Nic' (London, 1840), p. 11. It has been speculated that the author of the poem is Mary Martha Sherwood (1775–1851).

8   Alan Rauch, 'The Sins of Sloths: The Giant Ground Sloth as a Paleontological Parable', in *Victorian Animal Dreams*, ed. Deborah Morse and Martin Danahay (Aldershot, 2007), pp. 215–28.

9   Richard Owen, *Memoir on the Megatherium; or, Giant Ground-Sloth of America* (London, 1861).

10 Hesketh Prichard, *The Heart of Patagonia* (London, 1902), p. 361.

11 E. Ray Lankester, *Extinct Animals* (London, 1903), p. 184.

4 THE SIN OF SLOTH AND THE SLOTHS OF SIN

1   Horace, *The Odes and Satyrs of Horace: That Have Been Done Into English by the Most Eminent Hands* (London, 1721), p. 183.

2   Gonzalo Fernández de Oviedo, *The Natural History of the West Indies* [1526], trans. Sterling A. Stoudemire (Chapel Hill, NC, 1959), pp. 54–5.

3   Nehemiah Grew, *Musaeum Regalis Societati; or, A Catalogue and Description of the Natural and Artificial Rarities Belonging to the Royal Society and Preserved at Gresham Colledge* (London, 1681), p. 11.

4 Samuel Purchas, *Purchas His Pilgrimage; or, Relations of the World and the Religions Obserued in All Ages and Places Discouered*, 1st edn (London, 1613), p. 704.

5 Edward Bancroft, *Natural History of Guiana* (London, 1769), pp. 147–8.

6 George Shaw, *General Zoology* (London, 1800), vol. I, part 1, pp. 154–5.

7 Ibid., p. 155.

8 Alfred, Lord Tennyson, 'Lotos-Eaters' [1832], in *Tennyson's Poetry* (New York, 1971), p. 52.

9 St Thomas Aquinas, *Summa Theologica* [1485] (London, 1917), p. 463.

10 Dante Alighieri, *Purgatorio* [1472], Canto 18: ll. 130–41, trans. John D. Sinclair (Oxford, 1961), p. 239.

11 Geoffrey Chaucer and Larry Dean Benson, *The Riverside Chaucer* (Boston, MA, 1987), p. 730.

12 Charles Darwin, *On the Origin of Species* (London, 1859), p. 471.

13 John Bunyan, *The Pilgrim's Progress* [1678] (Ware, Herts, 1999), p. 6.

14 Samuel Johnson, *The Idler*, 1 (15 April 1758), p. 4, available at www.yalejohnson.com.

15 Herman Melville, *Bartleby the Scrivener* (New York, 1853).

16 Charles Dickens, *A Tale of Two Cities* [1859] (New York, 1960), p. 367.

17 Lewis Carroll, *Alice's Adventures in Wonderland* [1865] (New York, 1894), p. 96.

18 Angus Wilson, *The Seven Deadly Sins* [1962] (Pleasantville, NY, 2002), p. 58.

19 Yann Martel, *The Life of Pi* [2001] (New York, 2007), p. 5.

5 SLOTHS IN CULTURE

1 Robert S. Lemmon, *All About Strange Beasts of the Present* (New York, 1957), p. 88.

2 Juan Carlos Galeano, *Folktales of the Amazon* (Westport, CT, 2009), pp. 39–40.

3   Gaspar Morcote-Ríos et al., 'Colonisation and Early Peopling of the Colombian Amazon during the Late Pleistocene and the Early Holocene: New Evidence from La Serranía La Lindosa', *Quaternary International*, DLXXVIII (April 2021), pp. 5–19.

4   Charles Darwin, 'Bahia Blanca' (September–October 1832), *Geological Diary*, transcribed by Kees Rookmaaker, ed. John van Wyhe, p. 54, available at http://darwin-online.org.uk, CUL-DAR.63-72.

5   Robert FitzRoy, *Narrative of the Surveying Voyages of His Majesty's Ships Adventure and Beagle between the years 1826 and 1836. Proceedings of the Second Expedition, 1831–36, under the Command of Captain Robert Fitz-Roy, RN* (London, 1839), p. 107.

6   Claude Lévi-Strauss, *The Jealous Potter* (Chicago, IL, 1984), p. 152.

7   W. H. Hudson, *Green Mansions* [1904] (Oxford, 1998), p. 195.

8   William Gordon Stables, *In Quest of the Giant Sloth: A Tale of Adventure in South America* (London, 1902).

9   Ibid., p. 267.

10 Gonzalo Fernández de Oviedo, *Natural History of the West Indies* [1525], trans. Sterling A. Stoudemire (Chapel Hill, NC, 1959), p. 55.

11 Pierpont Morgan Library, *Histoire naturelle des Indes: The Drake Manuscript in the Pierpont Morgan Library* (New York, 1966), ff. 63v–64r.

12 Georges Louis Leclerc, Comte de Buffon, *Natural History*, ed. James Smith Barr (London, 1792), vol. VIII, p. 290.

13 Ibid., p. 289.

14 Simeon Shaw, *Nature Displayed*, Part IV (London, 1823), p. 383.

15 Charles Bell, *The Hand: Its Mechanism and Vital Endowments, as Evincing Design* (London, 1833), p. 27. This book was one of the *Bridgewater Treatises* commissioned to demonstrate the 'Power, Wisdom, and Goodness of God as Manifested in the Creation'.

16 Charles Darwin, 'Letter from Frederick William Hope' (15 January 1834), in *The Correspondence of Charles Darwin*, vol. I: *1821–1836* (Cambridge, 1985), p. 363.

17 William Buckland, *Geology and Mineralogy Considered with Reference to Natural Theology*, Bridgewater Treatises (London, 1836), p. 114.

18 Ibid., p. 130.

19 Gowan Dawson, 'Literary Megatheriums and Loose Baggy Monsters: Paleontology and the Victorian Novel', *Victorian Studies*, LIII/2 (Winter 2011), pp. 203–30.

20 Wayne Clough, 'A Giant Sloth Mystery Brought Me Home to Georgia', *Smithsonian Magazine*, www.smithsonianmag.com, 1 May 2019.

21 Richard Owen, *The Life of Richard Owen by His Grandson* (London, 1894), p. 190.

22 Charles Kingsley, *Alton Locke* (Oxford, 1983), p. 338.

23 Henry Morley, 'Our Phantom Ship on an Antediluvian Cruise', *Household Words* (London, 1851), pp. 492–6. See Benjamin Chandler's 'Some Guided Journey's to Early Times', *Palaeomedia*, https://palaeomedia.blogs.bristol.ac.uk, 16 February 2021.

24 William Makepeace Thackeray, *The Newcomes* (London, 1994), p. 474.

25 Charles Waterton, *Wanderings in South America* (London, 1836), p. 172.

26 Ibid., p. 169.

27 Henry Walter Bates, *The Naturalist on the River Amazons* (London, 1892), p. 207.

28 Elizabeth Agassiz, *A Journey in Brazil* (Boston, MA, 1869), p. 361.

29 William Beebe, 'The Three-Toed Sloth', *Zoologica*, VII/1 (March 1926), p. 3.

30 'Kristen Bell's Sloth Meltdown', *Ellen*, 31 January 2012, available at www.youtube.com.

31 'Programs', www.extremeanimals.org, accessed 12 August 2021.

32 Gemma Carter et al., 'The Impact of "Selfie" Tourism on the Behaviour and Welfare of Brown-Throated Three-Toed Sloths', *Animals*, VIII/11 (November 2018), p. 216.

33 Michael Levenson, '1,500 Eggs Were Waiting to Hatch. Then a Drone Crashed', *New York Times*, www.nytimes.com, 4 June 2021.

34 Jacqueline Jackson, *Julie's Secret Sloth* (Boston, MA, 1953).

35 Sylvaine Peyrols, *Le Paresseux* (Paris, 2013).
36 Lily Rothman, 'How Sloths Took Over Pop Culture, the World', *Time*, https://entertainment.time.com, 22 March 2013.
37 Gilbert Hernandez and Jared K. Fletcher, *Sloth* (New York, 2006).

6 SLOTHS, CONSERVATION AND ECO-AWARENESS

1 Jeremy Hance, 'Working to Save the "Living Dead" in the Atlantic Forest: An Interview with Antonio Rossano Mendes Pontes', *Mongabay*, www.mongabay.com, 23 September 2009.
2 John F. Eisenberg and Richard W. Thorington, 'A Preliminary Analysis of a Neotropical Mammal Fauna', *Biotropica*, v/3 (1973), pp. 150–61.
3 K. D. Neam, 'The Odd Couple: Interactions Between a Sloth and a Brown Jay', *Frontiers in Ecology and the Environment*, xiii/3 (April 2015), pp. 170–71.
4 Katherine Gammon, 'Animal Crossing: World's Biggest Wildlife Bridge Comes to California Highway', *The Guardian*, www.theguardian.com, 9 April 2022.
5 Mariana Simões, 'Brazil's Bolsonaro on the Environment, in His Own Words', *New York Times*, www.nytimes.com, 27 August 2019.
6 Tanya Dimitrova, 'Attempt to Export Nearly-Extinct Pygmy Sloths Sets off International Incident in Panama', *Mongabay*, www.mongabay.com, 20 September 2013.
7 Nisha Owen and Diorene J. Smith, 'Conserving the Pygmy Sloth', EDGE, https://ptes.org (June 2017).

Select Bibliography

Beebe, William, 'The Three-Toed Sloth', *Zoologica*, VII/1 (March 1926), pp. 1–67

Benson, Elizabeth P., *Birds and Beasts of Ancient Latin America* (Gainesville, FL, 1997)

Cliffe, R. N., et al., 'Mitigating the Squash Effect: Sloths Breathe Easily Upside Down', *Biology Letters*, X/4 (April 2014), article ID 20140172

Cliffe, Rebecca, and Suzi Esterhaus, *Sloths: Life in the Slow Lane* (Hayfield, Derbyshire, 2017)

Cooke, Lucy, *Life in the Sloth Lane: Slow Down and Smell the Hibiscus* (New York, 2018)

—, *The Truth about Animals: Stoned Sloths, Lovelorn Hippos, and Other Tales from the Wild Side of Wildlife* (New York, 2019)

Dawson, Gowan, 'Literary Megatheriums and Loose Baggy Monsters: Paleontology and the Victorian Novel', *Victorian Studies*, LIII/2 (Winter 2011), pp. 203–30

Fariña, Richard A., Gerardo De Iuliis and Sergio F. Vizcaíno, *Megafauna: Giant Beasts of Pleistocene South America* (Bloomington, IN, 2013)

Goffart, M., *Function and Form in the Sloth* (Oxford, 1971)

Lister, Adrian, *Darwin's Fossils: The Collection that Shaped the Theory of Evolution* (Washington, DC, 2018)

Martin, Paul S., *Twilight of the Mammoths: Ice Age Extinctions and the Rewilding of America* (Berkeley, CA, 2005)

Montgomery, G. Gene, *The Ecology of Arboreal Folivores: A Symposium Held at the Conservation and Research Center, National Zoological Park, Smithsonian Institution* (Washington, DC, 1978)

Morse, Deborah Denenholz, '"The Mark of the Beast": Animals as Sites of Imperial Encounter from *Black Beauty* to *Green Mansions*', in *Victorian Animal Dreams: Representations in Literature and Culture*, ed. Deborah Denenholz Morse and Martin Danahay (Aldershot, 2007)

Oviedo, Gonzalo Fernández de, *The Natural History of the West Indies* [1526], trans. Sterling A. Stoudemire (Chapel Hill, NC, 1959)

Peyrols, Sylvaine, *Le Paresseux* (Paris, 2013)

Prothero, David, *The Princeton Field Guide to Prehistoric Mammals* (Princeton, NJ, 2016)

Rauch, Alan, 'The Sins of Sloths: The Giant Ground Sloth as a Paleontological Parable', in *Victorian Animal Dreams*, ed. Morse and Danahay, pp. 215–28

Stables, Gordon, *In Quest of the Giant Sloth: A Tale of Adventure in South America* (London, 1902)

Trull, Sam, *Slothlove* (San Francisco, CA, 2016)

Vizcaíno, Sergio F., and W. J. Loughry, eds, *The Biology of the Xenarthra* (Gainesville, FL, 2008)

Waterton, Charles, *Wanderings in South America* (London, 1836)

Associations and Websites

AIUNAU FOUNDATION
www.aiunau.org

JAGUAR RESCUE CENTER
www.jaguarrescue.foundation
The Jaguar Rescue Center is a temporary or permanent home for
ill, injured and orphaned animals. With a focus on monkeys, sloths,
other mammals, birds and reptiles, the JRC provides veterinary
services, round-the-clock care and comfort to animals that would
otherwise be unable to survive in the rainforest or the sea of the
Caribbean.

SLOTH APPRECIATION SOCIETY
www.slothville.com
The Sloth Appreciation Society seeks to both protect the sloth
and promote the truth about its 'lazy' lifestyle. The much-maligned
sloth is an energy-saving icon with much to teach us humans about
sustainable living.

SLOTH CONSERVATION FOUNDATION (SLOCO)
https://slothconservation.org
SloCo was founded with the purpose of stimulating progressive
change and achieving lasting solutions through research and
conservation initiatives.

THE SLOTH INSTITUTE
www.theslothinstitute.org
The Sloth Institute, located in Manuel Antonio, Costa Rica, focuses on enhancing the welfare and conservation of sloths through the rescue, rehabilitation and release of hand-raised and injured sloths. While also conducting vital research, conservation and education programmes to ensure their survival. The Founder and Executive Director of the Institute is Sam Trull.

SLOTH SANCTUARY
www.slothsanctuary.com
The Sloth Sanctuary of Costa Rica is a privately owned animal rescue center located near the city of Cahuita. The sanctuary, founded by Judy Avey-Arroyo and Luis Arroyo, is dedicated to the rescue, rehabilitation, research, and release of injured or orphaned sloths. Tours of the sanctuary are offered to the public.

TOUCAN RESCUE RANCH
https://toucanrescueranch.org
The mission of the Toucan Rescue Ranch (TRR) is to rescue, rehabilitate and release Costa Rican wildlife. TRR has a model that focuses on conservation, education and research to ensure a brighter tomorrow for Costa Rican wildlife.

Acknowledgements

I am, foremost, grateful to Michael Leaman of Reaktion Books and to Jonathan Burt, editor of the Animal series, for their patience and guidance. This book was a long time in coming and both editors might well have imagined that it was actually being written *by a* sloth rather than *about* sloths. Two additional Reaktion editors, Alex Ciobanu and Phoebe Colley, have offered steadfast support and guidance that have made this book possible. I have benefited from my brief experience in Costa Rica, where I spent time at the Toucan Rescue Ranch (TRR). The director, Leslie Howle, was very supportive and included me in a variety of the sanctuary's activities. Other members of the TRR team were wonderful and helpful, and I am honoured to thank Carol Friesen, Janet Sandi Carmiol, Zara Palmer, Pedro Felipe Montero Castro and the rest of the team. Sam Trull, executive director of the Sloth Institute, is, to coin a phrase, a sloth dynamo, whose work, dedication and knowledge are truly an inspiration. I feel myself fortunate to have got to know her. Along with Sam, I must acknowledge both the images and the help I received from Erik F. Y. Hom and Maya Kaup of the University of Mississippi.

I am also fortunate to have worked with my student Carissa Wilbanks, who was an enormous help in the initial stages of *Sloth*. It is fitting that the final image of the book is her needlepoint image of a Bradypus, which is a gift I am happy to share with readers.

Many individuals have been generous with their time, their advice and even their illustrations. Kenny J. Travouillon, curator of Mammals at Western Australian Museum, tracked down a sloth in their collections to verify that sloths – like so many other animals he has described – fluoresce. Mark Witton, who combines scholarly expertise in palaeontology with superlative talent in illustration, not only provided images for the book but shared insights about sloth palaeobiology. The Smithsonian's Annette Aiello was happy to share her ground-breaking micrographs of sloth hairs. Brittney Elizabeth Stoneburg of the Western Science Center is responsible for wonderful photographs of fur from giant sloths. The cartoonist Michael Maslin was willing to tolerate a little taxonomic nit-picking on my part with humour (of course) and equanimity. Roger Hall's illustration of 'Sloths of the World' is an exceptional example of scientific illustration and I am very proud to have been able to include it in the book. I must also thank Dante Fabiero, the creator of Slothilda, for the use of his illustrations of a very imaginative sloth. The band The Sloths and artist Darren Merinuk were gracious enough to let me use the cover art from their album *Back from the Grave*, which I still listen to when getting ready to write. I have admired the brilliant music of Flanders and Swann since I was a child, and I am grateful to Leon Berger for proving permission to use the lyrics of 'Sloth'. Carl Buell's palaeontological illustrations have set a standard for excellence for decades, and I am grateful to him for his willingness to share his beautifully rendered images with both myself and the readers of this book.

Family and friends were a great support throughout, and they all tolerated more 'sloth talk' than any person deserves. My mother, Riva Rauch, and my sister, Joyce Rauch, listened patiently as I tried to explain why sloths are so captivating. Sadly, my mother died before she could see the final manuscript, but she took sloths to heart and supported this project (as did my sister) with love and enthusiasm. My friends and colleagues Juan Meneses and Allison Hutchcraft, both animal lovers, have supported both me and my sloth anecdotes while encouraging this project with camaraderie, insight, humour and wonderful poetry. Other friends and colleagues near and far, including Joanne Joy, Coren O'Hara, George Levine, Matt Rowney, Seirin Nagano, Kirk Melnikoff, Lara Vetter, Angie Williams and Itzik Pripstein, made the experience of writing this book and sharing its contents a genuine pleasure. My students, over the years, have patiently listened to sloth trivia and anecdotes without balking – but not without the occasional tongue-in-cheek sloth quip. The energetic presence of my companion dog, Kramer, who is anything but sloth-like, is always a joy, and a permanent reminder of the fascinating continuum between humans and all animals.

I owe a debt of gratitude to UNC Charlotte's College of Liberal Arts and Science for support in the acquisition of images and materials that helped make this book possible.

Finally, some nostalgic appreciation. The very first sloth that I encountered was the *Megatherium* that once dominated McGill's wonderful Redpath Museum. In my student days, studying vertebrate palaeontology under Robert Carroll, I learned from the often eccentric yet always brilliant Alice Johannsen, the director of the Redpath, that the museum's *Megatherium* was called simply 'Meg'. Eventually Meg left for a museum in Manitoba, but she still has a place in my heart.

The author and publishers wish to express their thanks to the sources listed below for permission to reproduce the following:

Page 138 from 'The Sloth', Trey Anastasio/Phish, *The Man Who Stepped into Yesterday* (1987). By permission of Redlight Management; page 139–9 from 'The Sloth', Michael Flanders and Donald Swann, *The Bestiary of Flanders and Swann*, Parlophone – PMC 1164 (1962). By permission of the Estates of Michael Flanders and Donald Swann. Administrator Leon Berger: leonberger@virginmedia.com.

Photo Acknowledgements

The author and publishers wish to express their thanks to the sources listed below for illustrative material and/or permission to reproduce it. Some locations of works are also given below, in the interest of brevity:

© 2002 20th Century Studios, Inc., all rights reserved: p. 147; Alamy Stock Photo: pp. 30 (*right*; PhotoStock-Israel), 31 (Hamza Khan), 76 (Chronicle); © 2022 Anderson Design Group, Inc., all rights reserved: p. 98; from *Annales du Muséum national d'histoire naturelle*, vol. v (Paris, 1804): p. 122; from Annette Aiello, 'Sloth hair: unanswered questions', in *The Evolution and Ecology of Armadillos, Sloths and Vermilinguas*, ed. G. Gene Montgomery (Washington, DC, 1985), reproduced courtesy of Annette Aiello: p. 58 (*left* and *right*); from *The Atlas of Nature: Being a Graphic Display of the Most Interesting Subjects in the Three Kingdoms of Nature* (London, 1823), photo collection of the author: p. 49 (*bottom*); from Josef Augusta, *Tiere der Urzeit* (Prague, 1960): p. 74; courtesy of Banco Central de Costa Rica (BCCR): p. 102; from Thomas Bewick, *A General History of Quadrupeds* (Newcastle upon Tyne, 1790): p. 118 (*centre*); courtesy of Carl Buell: pp. 61, 78; from *Canadian Illustrated News*, XXVI/10 (2 September 1882), photo McCord Stewart Museum, Montreal: p. 14 (*top*); courtesy of Eric Carle: p. 144; from *Cartoon Portraits and Biographical Sketches of Men of the Day* (London, 1873), photo collection of the author: p. 128; collection of the author: pp. 110, 121, 129, 151; from Frédéric Cuvier, *Dictionnaire des sciences naturelles* (Paris and Strasbourg, 1816–29): p. 30 (*left*); © 2016 Disney: p. 149; © Éditions Gallimard Jeunesse: pp. 57, 145; from George Edwards, *Gleanings of Natural History, Exhibiting Figures of Quadrupeds, Birds, Fishes, Insects . . .* , vol. II (London, 1760): p. 23; © Suzi Eszterhas/Minden Pictures: pp. 33, 60 (*centre*);

courtesy of Dante Fabiero: p. 152; photo © Thelmå Gatuźźo: p. 39 (*top*); George Eastman Museum, Rochester, NY: p. 43; © Roger Hall/inkart.net: p. 18; from *Harper's Weekly*, XXV/1273 (21 May 1881): p. 108; from *Illustrations of Zoology* (London and Glasgow, 1851): p. 117; courtesy of Professor José Iriarte: p. 104; from John Jonston, *Historia naturalis de quadrupetibus* (Frankfurt, *c*. 1653): p. 120 (*bottom*); from Maya Kaup, Sam Trull and Erik F. Y. Hom, 'On the Move: Sloths and Their Epibionts as Model Mobile Ecosystems', *Biological Reviews*, XCVI/6 (December 2021), images © 2021 The Authors (CC BY 4.0): pp. 22, 24, 32 (*left* and *right*); from Olivier Lambert, Giovanni Bianucci and Christian de Muizon, 'Macroraptorial sperm whales (Cetacea, Odontoceti, Physeteroidea) from the Miocene of Peru', *Zoological Journal of the Linnean Society*, CLXXIX/2 (2017): p. 62; photo © Pierre-Yves Le Bail: p. 40; from Georges-Louis Leclerc, Comte de Buffon, *Œuvres complètes de Buffon (avec la nomenclature Linnéenne et la classification de Cuvier)*, vol. III (Paris, 1853), photo collection of the author: p. 120 (*top*); McCord Stewart Museum, Montreal: p. 14 (*bottom*); McGill University Library, Rare Books and Special Collections, Montreal: p. 118 (*top*); The Metropolitan Museum of Art, New York: pp. 84, 87; The Morgan Library & Museum, New York (MA 3900, fol. 64r): p. 114; Museum of Fine Arts, Houston, TX: p. 115; Museum Wiesbaden (photo Klaus Rassinger and Gerhard Cammerer, CC BY-SA 3.0): p. 49 (*top*); The New Yorker Collection/The Cartoon Bank: pp. 25 (Michael Maslin), 96 (Tom Cheney); Zara Palmer/Toucan Rescue Ranch: pp. 17, 39 (*bottom*), 146; from *Pictures from Dickens with Readings* (London and New York, 1895): p. 92; from *Punch; or, The London Charivari*, vol. XIV (1848): p. 131; photos Alan Rauch: pp. 12, 35, 42 (*centre*), 48, 51, 53, 64, 66 (*left* and *right*), 161 (*right*); © Real Academia de la Historia, Madrid (M-RAH, 9/4786, fol. 166v): p. 86; © Joe Samson: p. 46; from *Scènes de la vie privée et publique des animaux*, vol. II (Paris, 1842), photo collection of the author: p. 67; Science Photo Library: pp. 69 (Paul D. Stewart), 72 (Mauricio Anton); from Simeon Shaw, *Nature Displayed in the Heavens, and on the Earth* (London, 1823), vol. IV: p. 119; Shutterstock.com: pp. 10 (Aldona Griskeviciene), 27 (*top*; Ged. I MacKenzie), 27 (*bottom*; Parkol), 28 (Marinaaaniram), 29 (Eduardo Menezes), 34 (Ondrej Prosicky), 38 (milan noga), 42 (*bottom*; Milosz Maslanka), 56 (Milan Zygmunt), 77 (Artush), 133 (Rob Jansen), 158 (Scenic Shutterbug), 160 (Katarzyna_Przygodzka);

courtesy of The Sloths and Darren Merinuk (the artist): p. 137; from J. G. Stedman, *Narrative of a Five Years' Expedition against the Revolted Negroes of Surinam*, vol. 1 (London, 1796): p. 124; from André Thevet, *Les singularitez de la France antarctique, autrement nommée Amérique* (Paris, 1558), photo collection of the author: p. 113; from *Thirty Plates Illustrative of Natural History* (London, 1842–50), photo collection of the author: p. 41; courtesy of Néstor Toledo: p. 81; courtesy of Dr Kenny J. Travouillon: p. 44; photos © Sam Trull, all rights reserved: pp. 6, 15, 60 (*top*), 156, 159, 161 (*left*); courtesy of James Tynion iv, Eryk Donovan and Adam Guzowski: p. 154; from Henry A. Ward, *Catalogue of Casts of Fossils, from the Principal Museums of Europe and America, with Short Descriptions and Illustrations* (Rochester, NY, 1866): p. 11; courtesy of Carissa Wilbanks: p. 162; courtesy of Mark Witton: p. 73 (*left* and *right*); © The WNET Group/ Karen Brazell: p. 59 (*bottom*).

Index

Page numbers in *italics* indicate illustrations